DEPOSITION, DIAGENESIS AND DOLOMITIZATION OF NEOGENE SEROE DOMI FORMATION CORAL REEF LIMESTONES ON CURAÇAO, NETHERLANDS ANTILLES

To My Family

DEPOSITION, DIAGENESIS AND DOLOMITIZATION OF NEOGENE SEROE DOMI FORMATION CORAL REEF LIMESTONES ON CURAÇAO, NETHERLANDS ANTILLES

with 93 text illustrations and 12 tables

by

Bruce W. Fouke

AMSTERDAM 1994

Uitgaven Natuurwetenschappelijke Studiekring voor het Caraïbisch Gebied: 133
Publications Foundation for Scientific Research in the Caribbean Region: 133

CIP-GEGEVENS KONINKLIJKE BIBLIOTHEEK, DEN HAAG

Fouke, Bruce W.

Deposition, diagenesis and dolomitization of neogene Seroe Domi formation coral reef limestones on Curaçao, Netherlands Antilles / by Bruce W. Fouke. – Amsterdam : Natuurwetenschappelijke Studiekring voor het Caraïbisch Gebied. – Ill., tab. – (Uitgaven Natuurwetenschappelijke Studiekring voor het Caraïbisch Gebied ; 133)
Met lit. opg.
ISBN 90-74624-03-0
Trefw.: koraalriffen ; Curaçao.

Copies can be purchased from:
Secretariat Foundation for Scientific Research
Inst. Tax. Zoology
Plantage Middenlaan 45
1018 DC Amsterdam, The Netherlands

TABLE OF CONTENTS

LIST OF FIGURES vii
LIST OF TABLES x
ABSTRACT xi
ACKNOWLEDGMENTS xiii

CHAPTER 1: INTRODUCTION 1
- Aim of the study 1
- Previous Geological Studies 2
- Regional Tectonic Setting 6
- Research Methods 11

CHAPTER 2: SEDIMENTOLOGY AND DEPOSITIONAL ENVIRONMENTS 15
- Introduction 15
- Lithofacies of the Seroe Domi Formation 15
  - Subunit 1 Lithofacies 1, 2, 3, and 4 19
  - Subunit 2 Lithofacies 5, 6, and 7 27
  - Subunit 3 Lithofacies 8 35
  - Tafelberg Phosphatization 35
- Quaternary Limestone Terraces 36
- Lithostratigraphic Distributions and Correlations 40
- Depositional Environments of the Seroe Domi Formation 41
  - Subunit 1 Depositional Environments 45
  - Subunit 2 Depositional Environments 47
  - Subunit 3 Depositional Environments 49
- Conclusions 50

CHAPTER 3: CHRONOSTRATIGRAPHY AND DEPOSITIONAL HISTORY 53
- Chronostratigraphy of the Seroe Domi Formation 53
  - Biostratigraphy 53
  - Sr Isotope Geochronology 56
- Depositional history of the Seroe Domi Formation 63
  - Subunit 1 63
  - Subunit 2 66
  - Subunit 3 68
- Geologic History of the Quaternary Terrace Limestones 68

Relative Sea Level History of Curaçao 69
Conclusions 73

CHAPTER 4: DOLOMITIZATION 75
Previous Studies of Seroe Domi Formation Dolomitization 75
Diagnostic Petrographic Criteria for Dolomites I, I' and II 77
Dolomite I and I' Petrography 79
Dolomite II Petrography 79
Distribution and Correlation of Dolomites I, I' and II 83
Geographic Distribution 86
Stratigraphic Distribution 87
Correlation of Dolomite Cathodoluminesence 94
Dolomite-Limestone Diagenetic Contacts 95
The Centripetal Dolomitization Replacement Process 96
Seroe Domi Formation Host Limestone Diagenesis 100
Pre-Dolomite I and I' Diagenesis 100
Pre-Dolomite II Diagenesis 104
Post-Dolomite I, I', and II Diagenesis 104
Geochemistry of Seroe Domi Formation Calcites and Dolomites 105
Skeletal Material and Pre- and Post-Dolomitization Calcites 105
Geochemistry of Dolomites I, I', and II 107
Spatial Distributions in Dolomite Geochemistry 111
Modern Curaçao Groundwaters 113
Geochemistry of Curaçao Lava Formation Groundwaters 115
Modelling of Fluid Saturation States and Mixtures 118
Interpretation of the Origin of the Curaçao Lava Formation Groundwaters 121
Evaluation of Dolomite Recrystallization 123
Reconstruction of Estimated Seawater Dolomite Compositions 124
Oxygen and Carbon Isotopes 124
Strontium Isotopes 126
Trace Element Abundances 126
Composition of Dolomite I, I', and II Dolomitizing Fluids 132
Stoichiometry 132
Strontium, Sodium, Fluorine, Chlorine, and Sulfate 132
Manganese and Iron 134
Potential Sources of Sr to the Dolomitizing Fluids 135
Chronology of Dolomite I, I', and II Precipitation 137
Timing of Dolomite I Precipitation 137
Timing of Dolomite I' Precipitation 139
Timing of Dolomite II Precipitation 140
Dolomitization History of the Seroe Domi Formation 141
Pre-Dolomite I and I' Diagenesis 142
Petrogenesis of Dolomite I 144
Petrogenesis of Dolomite I' 154
Pre-Dolomite II Diagenesis 159
Petrogenesis of Dolomite II 160
Post-Dolomite Diagenesis 165
Conclusions 165

REFERENCES 173

LIST OF FIGURES

Chapter 1

Figure 1.1. Geographic map of the Caribbean Sea. 3
Figure 1.2. Structural map of the Bonaire Block. 3
Figure 1.3. Geologic outcrop map of Curaçao. 4
Figure 1.4. Geologic succession of Curaçao. 5
Figure 1.5. Geologic setting of the Seroe Domi Formation. 6
Figure 1.6. Field photograph of the Seroe Domi Formation on Curaçao. 6
Figure 1.7. Geologic cross-section of present-day Curaçao. 7
Figure 1.8. Geographic map of Curaçao with sample localities. 9

Chapter 2

Figure 2.1. Geologic cross-section of southeastern Curaçao. 16
Figure 2.2. Composite stratigraphic column of the Seroe Domi Formation. 18
Figure 2.3. Sediment histograms of the Seroe Domi Formation. 19
Figure 2.4. Photograph of Subunit 1 type-section at Seru Blandan. 20
Figure 2.5. Line drawing of Seru Blandan type-section. 22
Figure 2.6. Line drawing of Seru Blandan type-section. 23
Figure 2.7. SEM images of Subunit 1 pelagic foraminifers. 25
Figure 2.8. SEM images of Subunit 1 benthic foraminifers. 26
Figure 2.9. Photograph of Subunit 2 type-section at Seru Mansinga. 28
Figure 2.10. Line drawing of Subunit 1 section at Seru Mansinga. 30
Figure 2.11. Photograph of Subunit 2 type-section at Seru Spreit. 31
Figure 2.12. Photograph of Subunit 2 and 3 section at St. Annabaai. 32
Figure 2.13. Line drawing Subunit 2 and 3 section at St. Annabaai. 34
Figure 2.14. Photograph of the Quaternary Limestone Terraces. 37
Figure 2.15. Lithostratigraphic diagram: Ser'i Kura to Duivelsklip. 38
Figure 2.16. Lithostratigraphic diagram: Evertzberg to Salina Abou. 38
Figure 2.17. Lithostratigraphic diagram: Seru Kabaye to Blauwbaai. 39
Figure 2.18. Lithostratigraphic diagram: Seru Zjeremi to Kabaye. 39
Figure 2.19. Geographic distributions of Subunits 1, 2, and 3. 41
Figure 2.20. Geologic map of Curaçao: Seru Kabritu to Blauwbaai. 42
Figure 2.21. Geologic map of Curaçao: Ser'i Kura to Ser'i Boka. 43

Chapter 3

Figure 3.1. Stratigraphic position of Subunit 1 foraminifer and oyster samples. 54
Figure 3.2. Miocene-Pliocene-Pleistocene stratigraphy. 55
Figure 3.3. Seroe Domi Formation coral fauna and biostratigraphy. 56
Figure 3.4. Stratigraphic position of Subunit 2 brachiopod and oyster samples. 59
Figure 3.5. Geochemical covariations in Subunit 1 oyster shells. 60
Figure 3.6. Shell material $^{87}Sr/^{86}Sr$ versus regression model ages. 61
Figure 3.7. Miocene-Pliocene depositional history of the Seroe Domi. 65
Figure 3.8. Pliocene-Pleistocene depositional history of the Seroe Domi. 67
Figure 3.9. Geologic cross-section of present-day Curaçao. 68
Figure 3.10. Summary of previous Quaternary Limestone analyses. 70
Figure 3.11. Summary of Seroe Domi Formation geochronology evidence. 71
Figure 3.12. Depositional history of the Seroe Domi Formation. 72

Chapter 4

Figure 4.1. Geologic cross-section of southeastern Curaçao. 76
Figure 4.2. Petrography of Dolomite I. 78
Figure 4.3. Petrography of Dolomite I'. 80
Figure 4.4. Petrography of Dolomite II. 82
Figure 4.5. Geographic map of Curaçao with sample localities. 85
Figure 4.6. Distributions of Dolomite I, I', and II on Curaçao. 87
Figure 4.7. Dolomite stratigraphy: Ser'i Kura to Duivelsklip. 88
Figure 4.8. Dolomite stratigraphy: Evertzberg to Salina Abou. 88
Figure 4.9. Dolomite stratigraphy: Seru Kabaye to Blauwbaai. 89
Figure 4.10. Dolomite stratigraphy: Seru Zjeremi to Seru Kabaye. 89
Figure 4.11. Dolomite I bodies at St. Michielsberg. 90
Figure 4.12. Correlation of Seru Blandan and St. Michielsberg. 91
Figure 4.13. Dolomite I stratiform distributions at Seru Blandan. 92
Figure 4.14. Dolomite I stratiform distributions at Seru Blandan. 93
Figure 4.15. Schematic diagram of SDB 1 and 2 at Seru Blandan. 96
Figure 4.16. Petrography of centripetal Dolomite I' replacement. 97
Figure 4.17. Petrography of centripetal Dolomite I replacement. 99
Figure 4.18. Seroe Domi Formation paragenetic sequence. 101
Figure 4.19. Pre-Dolomite I and I' leaching and calcite cementation. 102
Figure 4.20. Pre-Dolomite I and I' bladed calcite cements. 103
Figure 4.21. Dolomite, calcite and skeletal $\delta^{18}O$ versus $\delta^{13}C$. 105
Figure 4.22. SXRFMA spectra from Dolomites I, I', and II. 110
Figure 4.23. Regional variations in dolomite Mn, Sr, and Na. 112
Figure 4.24. Regional variations in dolomite Cl, F, and SO_4. 113
Figure 4.25. Regional variations in dolomite mol% Ca and $^{87}Sr/^{86}Sr$. 114
Figure 4.26. Regional variations in dolomite $\delta^{18}O$ versus $\delta^{13}C$. 115
Figure 4.27. Geochemical composition of Curaçao groundwaters. 117
Figure 4.28. Saturation indices for Curaçao groundwaters. 119
Figure 4.29. Saturation indices for mixed Curaçao groundwaters. 120
Figure 4.30. Correlations of geologic and geochemical anomalies. 123
Figure 4.31. Dolomite $\delta^{18}O$ versus $\delta^{13}C$ compositions. 125
Figure 4.32. Seroe Domi and Tertiary Sr and Na abundances. 128
Figure 4.33. Comparison of Na versus Cl in Dolomites I, I', and II. 130
Figure 4.34. Dolomite SO_4 abundance versus excess Na and Cl. 131
Figure 4.35. Sources of Sr on Curaçao. 136
Figure 4.36. Dolomite I, I', and II $^{87}Sr/^{86}Sr$ model ages. 138
Figure 4.37. Seroe Domi stratigraphic and geochemical evidence. 141
Figure 4.38. Depositional and diagenetic history of the Seroe Domi. 142
Figure 4.39. Seroe Domi calcite and shell $\delta^{18}O$ and $\delta^{13}C$. 144
Figure 4.40. Dolomite I $\delta^{18}O$ versus $\delta^{13}C$ evaluation. 146
Figure 4.41. Dolomite I $^{87}Sr/^{86}Sr$ versus $\delta^{18}O$ and $\delta^{13}C$. 147
Figure 4.42. Dolomites I, I', and II $^{87}Sr/^{86}Sr$ versus Sr abundances. 148
Figure 4.43. Covariations in Dolomite I Na, SO_4, Cl, Sr, and Mn. 149
Figure 4.44. Dolomite I, I', and II $^{87}Sr/^{86}Sr$, ^{13}C and excess Na. 151
Figure 4.45. Saturation indices for groundwater mixing with seawater. 152
Figure 4.46. Dolomite I' $\delta^{18}O$ versus $\delta^{13}C$ evaluation. 155
Figure 4.47. Covariations in Dolomite I' $^{87}Sr/^{86}Sr$, $\delta^{18}O$ and $\delta^{13}C$. 157

Figure 4.48. Covariations in Dolomite I' Na, SO_4, Cl, Sr, and Mn. 158
Figure 4.49. Variations in $^{87}Sr/^{86}Sr$ during freshwater-seawater mixing. 161
Figure 4.50. Dolomite II $\delta^{18}O$ versus $\delta^{13}C$ evaluation. 163
Figure 4.51. Dolomite II $^{87}Sr/^{86}Sr$ versus $\delta^{18}O$ and $\delta^{13}C$. 164
Figure 4.52. Dolomite II Na, SO_4, Cl, Sr, and Mn compositions. 166

LIST OF TABLES

Chapter 1

Table 1.1. Locality names of individual Seroe Domi Formation exposures. 8

Chapter 2

Table 2.1. Sedimentologic summary of the Seroe Domi Formation. 16
Table 2.2. Paleontological assemblages observed in the Seroe Domi Formation. 17

Chapter 3

Table 3.1. Geochemistry of Recent and Seroe Domi Formation shell samples. 58
Table 3.2. Sr isotope model ages of Seroe Domi Formation shell samples. 62

Chapter 4

Table 4.1. Seroe Domi Formation Dolomite I, I', and II distributions. 84
Table 4.2. Geochemistry of calcite cements and whole rock analyses. 106
Table 4.3. Geochemistry of Seroe Domi Dolomites I, I', and II. 107
Table 4.4. Geochemistry of Dolomite I, I', and II and the ESD. 108
Table 4.5. Modern Curaçao groundwater compositions. 116
Table 4.6. Geochemical compositions of natural near-surface waters. 121
Table 4.7. $^{87}Sr/^{86}Sr$ model ages for Dolomites I, I', and II. 139

ABSTRACT

The Seroe Domi Formation is an extensively dolomitized 350 m-thick sequence of limestones that stand in relief as a dissected mountain belt on the leeward coast of Curaçao, Netherlands Antilles. The formation is composed of three stratigraphic subunits and eight lithofacies that were deposited on the flanks of the Curaçao horst in response to hundreds of meters of eustatically and tectonically induced relative sea level changes. Field relationships, sedimentologic analyses, foraminifer and coral biostratigraphy, and skeletal Sr chronostratigraphy have been combined to establish the depositional history of the Seroe Domi Formation. Subunit 1 is composed of interbedded shelf-derived carbonate gravity flows and pelagic oozes deposited during the Middle Miocene in deep-water (>500 m) environments. Subunit 2 is composed of siliciclastic-rich sands derived from the Curaçao basement and deposited in moderately deep-water environments during the Late Miocene to Early Pliocene. The coralgal grainstones and boundstones comprising uppermost Subunit 2 were deposited in shallow-water (<100 m) environments during the Late Miocene and Pliocene. And finally, Subunit 3 basalt-derived siliciclastic sands were deposited within Subunit 2 erosional voids during the Early Pleistocene.

Three petrographically and geochemically distinct types of dolomite have been observed within the Seroe Domi Formation, herein called Dolomites I, I', and II. Covariations in dolomite isotopic and trace element compositions have been evaluated with respect to the host limestone paragenetic sequence, stratigraphic distributions of the dolomite samples, and the tectonic-eustatic relative sea level history of Curaçao. These analyses suggest that Dolomites I, I', and II were precipitated from fluids derived from the mixing of normal seawater with isotopically depleted freshwater. Sr isotope geochronology suggests that Dolomite I was precipitated during the late

Middle Miocene to early Late Miocene, Dolomite I' was precipitated during the Late Miocene to Early Pliocene, and Dolomite II precipitation took place during the Late Pliocene to Early Pleistocene. Unlike Dolomites I and I', Dolomite II exhibits wide absolute variations in $^{87}Sr/^{86}Sr$ that range from being equivalent, to being significantly *less* radiogenic than the $^{87}Sr/^{86}Sr$ of skeletal material in the host limestones. In addition, Dolomite II is significantly enriched in Zn and Cu. Analogy with modern Curaçao Lava Formation groundwaters suggests that Dolomite II was precipitated from fluids created by the mixing of normal seawater with meteoric groundwater that had derived Zn, Cu, and Sr from water-rock interaction with weathered basalts and basalt-derived siliciclastics. Petrographic relationships and Sr data suggest that each individual episode of dolomitization took place soon after subaerial exposure and stabilization of the host limestones, and correlate well with eustatic third-order sealevel lowstands and their ensuing sea level rises.

ACKNOWLEDGMENTS

This research was completed as part of the doctoral dissertation of Bruce Fouke at the State University of New York at Stony Brook under the direction of William Meyers. Gilbert Hanson, Richard Reeder, Scott McLennan, Robert Aller, and Philip Choquette all made significant contributions to this work. Dirk Beets reviewed the manuscript. Financial support for this research was provided by United States Department of Energy contract DE-AC02-83ER13112 held by G. Hanson and W. Meyers. Additional support was provided by The American Association of Petroleum Geologists, The Geological Society of America, Texaco Oil Company, Exxon Production Company, Shell Oil Company, Mobile Oil Company, and the Cave Research Foundation.

The geochemical and biostratigraphic analyses conducted in this research required the skills and collaboration of many people, with special thanks to the following: C.J. Beets for conducting Sr isotope analyses at the Vrije Universiteit, W. Staudt for assistance in IC analyses at Stony Brook, W. Dawson and Texaco Oil Company for analyzing O and C isotopes, P.C. Smalley and British Petroleum for conducting O and C isotope laser ablation analyses, B. Ward for assistance in conducting synchrotron analyses at Brookhaven National Laboratory, C. Winkel and the Island Government of Curaçao for assistance in groundwater sampling and providing unpublished groundwater analyses, R. Stoessel for modelling the saturation states of modern groundwaters; S. McLennan for assistance in operating the DCP at Stony Brook, R. Reeder for completing TEM analyses and assistance in operating the XRD at Stony Brook, A. Melillo and Chevron U.S.A. Production Company for completing the foraminifer identifications, and A. Budd for completing the coral identifications at the University of Iowa.

Geologists familiar with the Netherlands Antilles provided invaluable help during the initiation of this project, and I would like to thank D. Sibley, D.J. Beets, G.Th. Klaver, P. Stienstra, T. ten Have, R. Murray, R.P.M. Bak, C. Schubert, and D. Kobluk. I also owe much for the hospitality and assistance provided by the people of Curaçao, Bonaire and Aruba, during the course of the field work. *Masha Danki!* I would like to give particular thanks to C.W. Winkel, E.N. Ayubi, J.B. Haviser, the entire staff at the Caribbean Marine Biological Institute (CARMABI), W.P. Maal, A. Kooij, C. Racamy, H. Coffei, R. Alderlieste, Mr. and Mrs. P. Eddine, E.A. Curet, R. Brown, Mrs. Hoevert, R. van Halewijn, E. Newton, J. Braafhart, J. Gielen, and C.F. Reudink.

The successful completion of this study was due solely to the help and encouragement of my family, to whom I wish to express my sincere appreciation! My deepest thanks to my wife Ann, who assisted in the field work, typed, edited, drafted, completed dark room work, and tirelessly supported me in every aspect of this project.

CHAPTER 1

INTRODUCTION

Aim of the study

Dolomitization is the name given to the geological process by which a limestone originally composed of calcium carbonate ($CaCO_3$) is partially or completely transformed into the mineral dolomite ($CaMg(CO_3)_2$). This transformation takes place as the combined result of two physical processes. The first is the direct precipitation of dolomite cement from a fluid into void spaces in the limestone, while the second is the replacement of the original limestone calcite with dolomite via solution and reprecipitation. The majority of limestones within the crust of the earth have been dolomitized, and many of these are associated with economically important reserves of hydrocarbons and ore deposits. Yet despite concentrated research, many aspects of our knowledge about the dolomitization process remain incomplete. Several recent studies have been able to reconstruct the composition and timing of the fluids responsible for the dolomitization of ancient limestones reasonably well (e.g. Choquette and Steinen, 1980; Aahron *et al.*, 1987; Vahrenkamp; 1988; Wilson *et al.*, 1989; Banner *et al.*, 1989). However, further progress in understanding how diagenetic waters dolomitize carbonate minerals requires that partially dolomitized limestones be chosen for study that exhibit the intermediate as well as endmember stages of the dolomitization process.

With these considerations in mind, the Seroe Domi Formation limestones on Curaçao have been chosen as a natural laboratory in which to study the process of dolomitization The Seroe Domi Formation exhibits several unique attributes that have permitted characterization of the incipient stages of near-surface limestone dolomitization, including: 1) non-dolomitized, partially dolomitized, and completely dolomitized limestones; 2) contacts between pervasively dolomitized limestones and non-dolomitized limestones that contain the intermediate steps of the dolomitization process; 3) dolomite fabrics and distributions directly analogous to older Phanerozoic dolomites; 4) excellent exposure of the formation as a 40 km-long series of dissected mountain ridges permitting centimeter-scale three-dimensional mapping of dolomite and the host limestone lithologies; and 5) deposition on the flanks of an uplifted basaltic horst, therein precluding burial and limiting the potential complexity and variety of dolomitizing fluids.

The stratigraphy and depositional history of the Seroe Domi Formation has been established in this study in order to provide a spatial and temporal framework within which to reconstruct the composition, timing, and distribution of the dolomitizing fluids. Previous work on the Seroe Domi Formation provided a geologic basis for the present study (Deffeyes *et al.*, 1965; Murray, 1969; De Buisonjé, 1974; Bandoian and Murray, 1974; Sibley, 1980, 1982; Stienstra, 1983, 1991; Ten Have *et al.*, 1984). The research presented in this study is organized into four chapters, each of which play an integral part in establishing the depositional and diagenetic history of the Seroe Domi Formation on Curaçao. Previous studies, the regional tectonic setting, and research methods are summarized here in Chapter 1. Chapter 2 establishes the sedimentology, stratigraphy, and depositional environments of the Seroe Domi Formation limestones. Chapter 3 synthesizes the lithostratigraphy presented in Chapter 2 with biostratigraphic and chemostratigraphic techniques to determine the depositional history of the Seroe Domi Formation. And Chapter 4 utilizes petrographic and geochemical analyses of dolomites, calcites and skeletal material to reconstruct the dolomitization history of the Seroe Domi Formation.

Previous Geological Studies of Curaçao and the Seroe Domi Formation

The east-west trending Aruba-La Blanquilla island archipelago lies 20 to 150 km off the coast of Venezuela at the southern margin of the Caribbean Basin (Fig. 1.1). Collectively referred to as the Leeward Lesser Antilles (Fig. 1.2), these islands include Aruba, the Netherlands Antilles (Curaçao and Bonaire) and the Venezuelan Antilles (Las Aves, Los Roques, La Orchila, and La Blanquilla). Curaçao (12° 2' N to 12° 23' N Latitude, 68° 44' W to

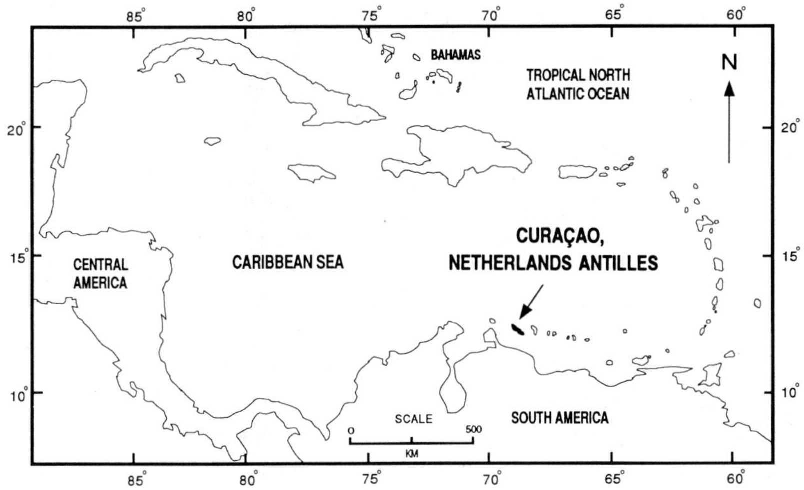

FIGURE 1.1. Geographic location of Curaçao, Netherlands Antilles, within the Caribbean Sea.

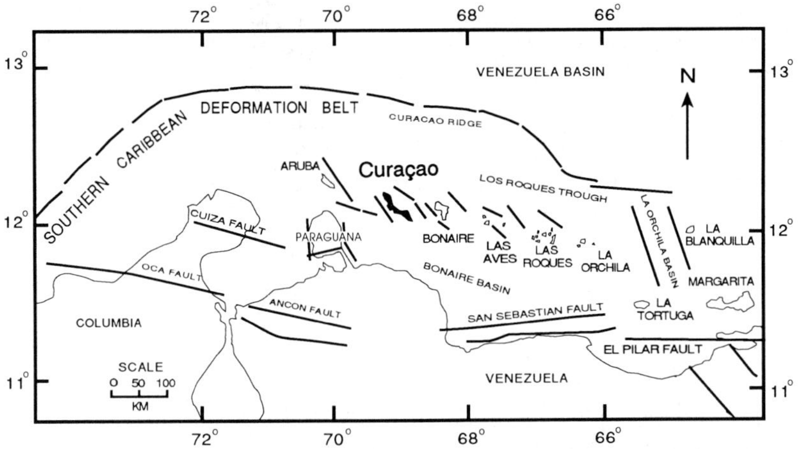

FIGURE 1.2. Structural map of the northwestern Venezuela borderland and the Bonaire Block (after Muessig, 1984).

68° 10' W Longitude), the largest of the islands, is 61 km in length by 7 to 14 km in width, with an emergent surface area of approximately 475 km^2 (Fig. 1.3).

The island of Curaçao is composed of a complex succession of Mesozoic through Cenozoic igneous and sedimentary lithologies (Fig. 1.4). The base-

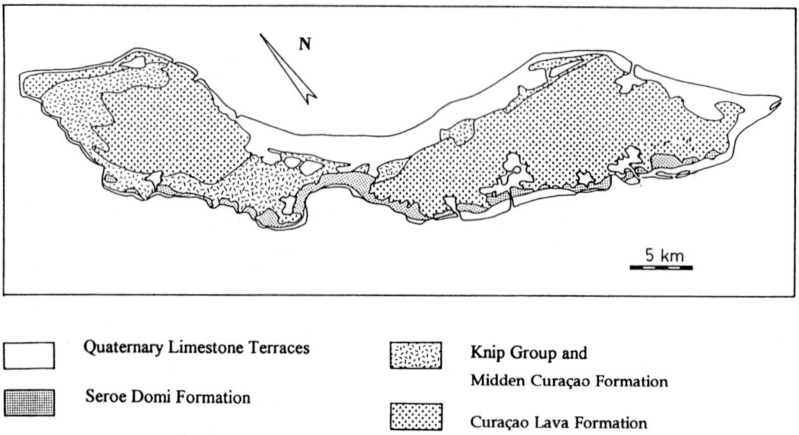

FIGURE 1.3. Geologic outcrop map of Curaçao (from Stienstra, 1991, which was modified from Beets, 1972).

ment rocks of Curaçao have been compressed, folded and faulted, forming two large anticlinoria separated by a central synclinorium (Beets, 1972; 1974; Fig. 1.3). The southeastern and northwestern anticlines are broad dome-like structures composed primarily of Cretaceous basalts, while the central syncline is composed of Cretaceous through early Tertiary metavolcanic and metasedimentary deposits (Beets, 1972; Klaver, 1987). The Curaçao basement is unconformably overlain by Eocene, Miocene, Pliocene and Pleistocene sedimentary deposits (Beets, 1972; De Buisonjé, 1974; Fig. 1.4). Eocene sediments are limited to four scattered outcrops of limestone, marl and clay, the thickest of which is at Ser'i de Cueba (Schaub, 1948; MacGillavry, 1977; Beets, 1972). Early Miocene deposition is confined to a thin 1 m-thick layer of limonitic packstones at the base of Tafelberg mountain, which contain *Miogypsina globulina* wackestone nodules (De Buisonjé, 1974).

The Neogene Seroe Domi Formation, a 350 m-thick sequence of carbonate wackestones, packstones, grainstones and siliciclastics, stands in relief along the leeward margin of Curaçao as inclined ridges onlapped by Quaternary Limestone Terraces (Figs. 1.3, 1.5, 1.6 and 1.7). De Buisonjé (1974) observed interbedded carbonate debris-flows, volcaniclastics, and hemipelagic slope sediments in the Seroe Domi Formation, and interpreted the coral, bivalve and echinoid assemblages to be Late Miocene to Pliocene in age. The inclined bedding and leeward distribution of the Seroe Domi

PERIOD	EPOCH	CURAÇAO GEOLOGY
TERTIARY	Pleistocene	Quaternary Limestone Terraces Age: Pleistocene > 50 m of coralgal limestones
TERTIARY	Pliocene Miocene	**Seroe Domi Formation** **Age: Miocene, Pliocene, Pleistocene** **350 m of mixed coralgal limestones and siliciclastic sandstones**
TERTIARY	Miocene	Early Miocene foraminifer clays and marls
TERTIARY	Oligocene	Hiatus
TERTIARY	Eocene	Ser'i de Cueba Limestones Age: Eocene 20 m of fossiliferous limestones
TERTIARY	Eocene / Paleocene	Hiatus
TERTIARY / CRETACEOUS	Paleocene / Late	Midden Curaçao Formation Age: Maastrichtian to Paleocene 1000 m of siliciclastic turbidites, slump breccias and shales
CRETACEOUS	Late	Knip Group Age: Campanian to Maastrichtian 2000 m of siliciclastic turbidites, siliceous pelagic sediments and tuffs Seroe Teintje Limestones
CRETACEOUS	Late	Zevenbergen and Casabao Fm. Limestones
CRETACEOUS	Late	Hiatus
CRETACEOUS	Early	Curaçao Lava Formation Age: Albian 5000 m of pillow basalts and diabase sills

FIGURE 1.4. Stratigraphic succession of Curaçao compiled from Beets (1972), De Buisonjé (1974), Beets *et al.* (1984), Klaver (1987), and the present study.

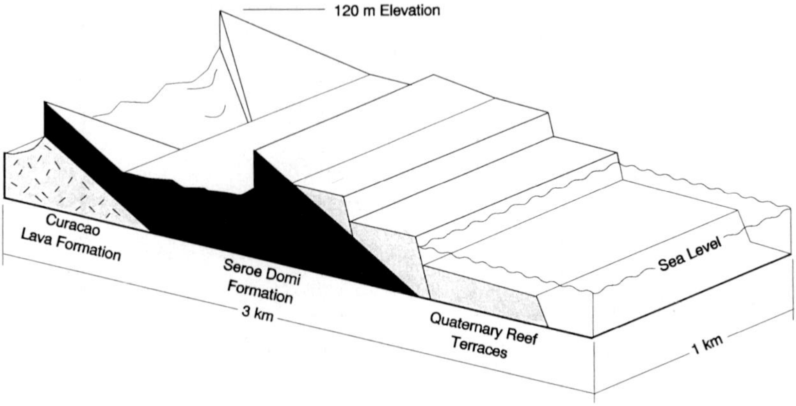

FIGURE 1.5. Geologic setting of the Seroe Domi Formation on Curaçao. Three-dimensional geologic block diagram of the Seroe Domi Formation on the leeward coast of Curaçao. The Seroe Domi Formation unconformably onlaps the preferentially eroded Curaçao Lava Formation, and now stands in relief as inclined flatiron ridges onlapped by the Quaternary Limestone Terraces. Modified from De Buisonjé (1974).

FIGURE 1.6. Photograph of the Seroe Domi Formation outcrop belt on the leeward coast of Curaçao. Photograph taken looking toward the southeast from St. Michielsberg.

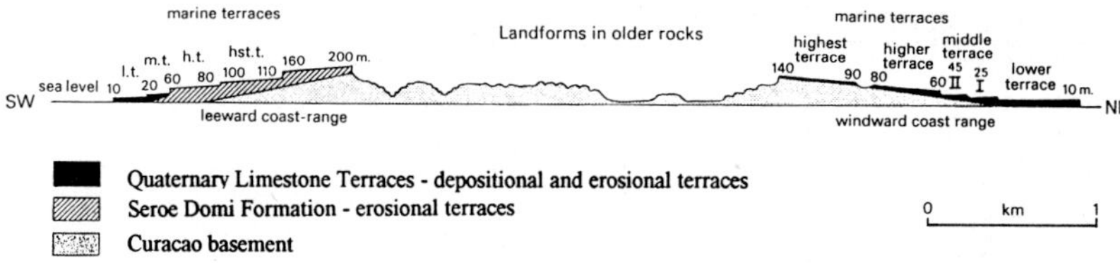

FIGURE 1.7. Geologic cross-section of the modern island of Curaçao. Cross-section illustrates the elevation of depositional and erosional terraces of the Seroe Domi Formation and Quaternary Limestone Terraces. From Herweijer (1979), which was modified from De Buisonjé (1974).

Formation was interpreted as original leeward depositional slope in deep forereef marine environments, based on: 1) *in situ Lithophagus* geopetals; 2) strike and dip parallel to the present-day shelf and coastline; 3) present-day easterly wind, wave and current motion; 4) poorly sorted sediments with no *in situ* corals; and mixed shallow and deep water faunal assemblages (De Buisonjé, 1974).

The lithofacies, stratigraphy and radiometric ages of the aerially extensive Quaternary Limestone Terraces that unconformably onlap the Seroe Domi Formation (Figs. 1.5 and 1.7) have been described by Alexander (1961), Murray (1969), De Buisonjé (1974), Bandoian and Murray (1974), Herweijer and Focke (1978), and Schubert and Szabo (1978). De Buisonjé (1974) observed 5 terrace levels, which he named according to their relative elevation as: Highest Terrace, Higher Terrace, Middle Terrace II, Middle Terrace I, and Lower Terrace. Herweijer and Focke (1978) subsequently found that these major terrace units are composed of 10 depositional subunits, the oldest of which is at least 470,000 years in age. Each of these subunits is capped with truncation surfaces and caliches depleted in $\delta^{13}C$ (Herweijer and Focke, 1978).

Regional Tectonic Setting

The Venezuelan Borderland, which extends for over 2,000 km northward from the northern portion of South America into the Caribbean Sea, is a tectonically complex area of depositional basins and submarine ridges that are dissected by large transform and normal fault systems (Silver *et al.*, 1975; Muessig, 1984). Curaçao lies within the region called the Bonaire Block, a piece of stretched continental crust lying between the Caribbean and South

LOCALITY NAME	SUBUNIT	LITHOFACIES
1) Seru Belefa	2	7
2) Knipberg	2	7
3) Seru Zjeremi	2, 3	7, 8
4) Ser'i Baha So	2	7
5) Ser'i Kalki	2	7
6) Seru Bo Kuki	2	7
7) Ser'i Boka	2	7
8) Seru Grandi	2	7
9) Seru Largu	2	7
10) Seru Lakete	2	7
11) Seru We'i Awa	2	7
12) Seru Nobo	2	7
13) Seru Kabaye	2, 3	7, 8
14) Seru Kabritu	2	7
15) Seru Rinkon	2	7
16) Seru Lucia	2	7
17) Seru Largu	2	7
18) Seru Popchi	2	7
19) Seru Hermanus	2	7
20) Ser'i Mahoc	2	7
21) Seru Pais	2	7
22) Ser'i Kabritu	1, 2	1, 6, 7
23) Seru Sami	1, 2	1, 6, 7
24) Seru Spreit	1, 2	1, 6, 7
25) Sint Michielsberg	1, 2	1, 2, 3, 6, 7
26) Seru Blandan	1, 2	1, 2, 3, 6, 7
27) Blauwbaai	2	7
28) Jack Evertszberg	1, 2	1, 6, 7
29) Veerisberg	1, 2	1, 6, 7
30) Seru Pretu	1, 2	1, 6, 7
31) Planters Rust	2, 3	7, 8
32) Ser'i Domi	1, 2	1, 3, 6, 7
33) Sint Annabaai	2, 3	6, 7, 8
34) Fort Nassau	1, 2, 3	1, 6, 7, 8
35) Ser'i Salina Abou	2	7
36) Ser'i Kura	1, 2	1, 6, 7
37) Seru Spano	2	7
38) Seru Mansinga	1	5
39) Kabrietenberg	1, 2	1, 4, 6, 7
40) Ser'i Boka	1, 2	1, 4, 6, 7
41) Tafelberg	1?	1?, 4, +?
42) Seru Magasina	1	1, 4
43) Seru Prekstul	1	1, 4
44) Duivelsklip	1	4

TABLE 1.1. Locality names of individual flatiron exposures of the Seroe Domi Formation on Curaçao, Netherlands Antilles. Numbers refer to localities shown in Figure 1.8.

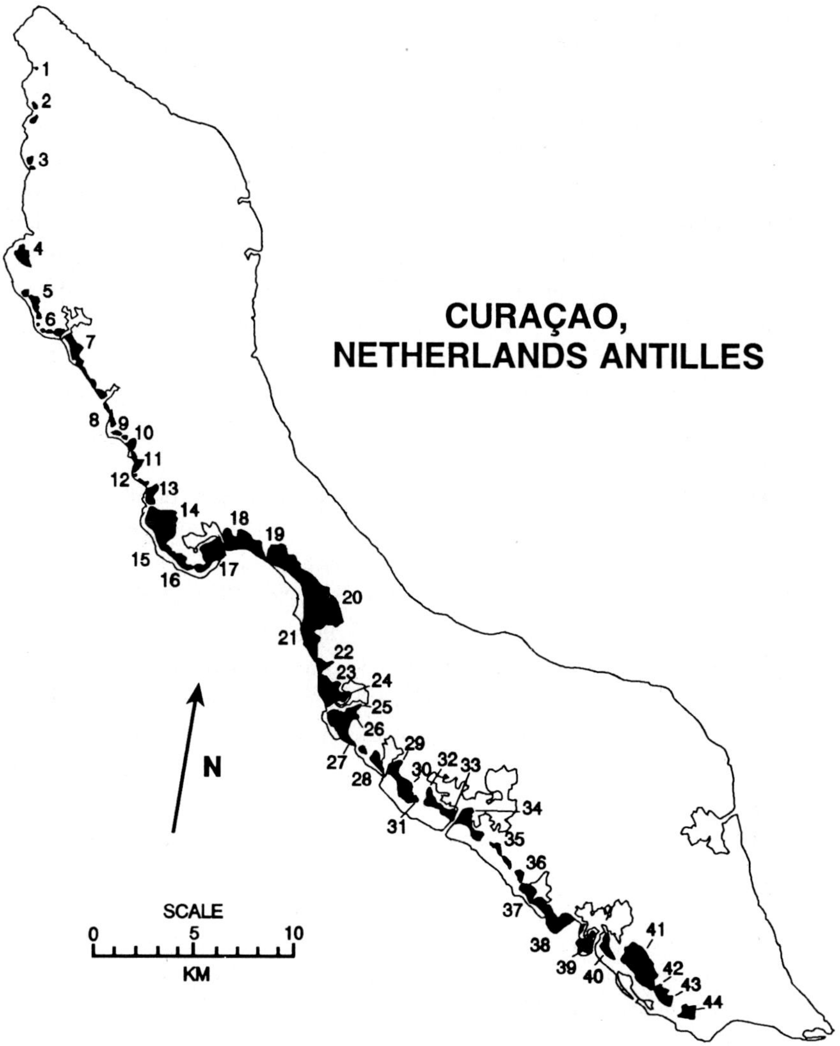

FIGURE 1.8. Geographic map of the island of Curaçao, Netherlands Antilles. The Seroe Domi Formation outcrop belt on the leeward coast of Curaçao is in black (after De Buisonjé, 1974), and numbered outcrops refer to locality names presented in Table 1.1. Each numbered outcrop represents a locality where the Seroe Domi Formation section was measured and sampled.

American plates (Fig. 1.2). The Bonaire Block is delimited by the Southern Caribbean Deformed Belt (also called the Curaçao Ridge) to the north and west, the La Orchila Basin (also called the Los Roques Canyon) to the east, and the Oca-Cuiza-San Sebastian fault zone to the south (Case, 1974; Silver *et al.*, 1975; Beets *et al.*, 1984; Muessig, 1984; Mann *et al.*, 1990; Fig. 1.2). The Leeward Lesser Antilles are the emergent portions of a thick submarine ridge composed of basalts and metasediments that parallel the deformed strata of the Los Roques Trough and the Curaçao Ridge (Silver *et al.*, 1975; Beets *et al.*, 1984). Curaçao, and each of the other Leeward Lesser Antilles, is bounded by northwest-trending normal faults that strike parallel to the length of the island at approximately N130°E, creating a linear series of horsts separated by 1 to 2 km-thick sediment filled grabens (Case and Holcombe, 1980; Silver *et al.*, 1975; Muessig, 1984). To the south of the islands, the 2 to 5 km-thick deposits of Tertiary clastics in the Bonaire Basin also contain north to northwest trending normal faults, with the rhombic form of these basins further suggesting that they are subsiding graben structures surrounded by northwest and east-west oriented normal faults (Case and Holcombe, 1980; Silver *et al.*, 1975; Muessig, 1984; Fig. 1.2).

Although details of the process are still highly controversial, it is generally accepted that the Caribbean plate has been converging in an east-southeasterly direction on the northern margin of the South American plate since the Oligocene (Hess and Maxwell, 1953; Maresch, 1974; Santamaria and Schubert, 1974; Kellog and Bonini, 1982; Pindell and Dewey, 1982; Perez and Aggarwal, 1991). The right-lateral shearing generated by this oblique convergence has created parallel zones of pull-apart basins and push-up blocks in the Venezuelan Borderland via crustal extension and subsidence developed at bends in the transcurrent fault system (Muessig, 1984). The Southern Caribbean Deformed Belt is interpreted to be an accretionary wedge formed by Miocene subduction of the Caribbean crust, which has been folded, faulted and overthrust to the northwest while accomodating these regional transform and convergent plate motions (Biju-Duval *et al.*, 1982; Mann *et al.*, 1990).

Vertical displacements of the Curaçao horst during this complex history of regional tectonism was presumably most intense during the Cretaceous to Eocene (Silver *et al.* 1975; Pindell and Dewey, 1982), resulting in more than 5 km of uplift and compressional folding (Beets, 1972; Klaver, 1987). The ensuing Neogene is assumed to have been a tectonically quiescent period of gradual uplift based on the measurement and dating of planation surfaces and coral reef terraces in the Leeward Lesser Antilles (Graf, 1972; Vignali, 1972; Schubert and Valastro, 1976; Schubert *et al.*, 1977; Herweijer and Focke, 1978; Schubert and Szabo, 1978; Mann *et al.*, 1990). However, stratigraphic evidence compiled in the present study will be presented that suggests substantial vertical displacements of the Curaçao horst took place as recently as the Middle Miocene.

Research Methods

Sixty sections from 42 localities on Curaçao have been measured and sampled during 9 months of field work (Table 1.1, Fig. 1.8). Geologic maps by Beets (1972) and De Buisonjé (1974), and 1:25,000 scale topographic maps (Cadastral Survey Department of the Netherlands Antilles, 1982) served as base maps for this study. For consistency in stratigraphic nomenclature and geography, the word "Seroe" was retained for formation names (after Westermann, 1935), while the words "Seru" or "Ser'i" were used for the names of individual mountains as presented in the 1982 topographic maps prepared by the Netherlands Antilles Cadastral Survey.

Pelagic foraminifers were identified by A.J. Melillo at Chevron USA in New Orleans. Corals were identified by A. Budd at the University of Iowa. Oyster and brachiopod shells, calcite cements, and dolomitized limestones chosen for chemical analyses were petrographically examined using plane-light, cathodoluminesence (Technosyn Luminoscope operated at 10 to 12 kV) and SEM (JEOL JSM 25S operated at 15 kV) microscopy, and micro-drilled using dental equipment. Each calcite and dolomite microsample powder was ground to 75 m mesh, and evaluated for mineralogical homogeneity by x-ray diffractometry (XRD) on a SCINTAG XRD at Stony Brook. Dolomite powders exhibiting an XRD calcite peak were leached in 4% acetic acid, rinsed in deionized water, and x-rayed again until calcite phases could no longer be detected (Staudt *et al.*, 1993). This produced 99% pure dolomite powder splits that were used for DCP, IC and mass spectrometry analyses.

$^{87}Sr/^{86}Sr$ ratios were measured by C.J. Beets on a Finnigan MAT 261 fixed multicollector thermal ionization mass spectrometer at the Isotope Geology Laboratory of the Free University in Amsterdam, The Netherlands. Analytical techniques are presented in Beets (1991; 1992). Sr isotope ratios from 10 replicate analyses of NBS SRM 987 yielded a mean value of 0.710258 with a standard deviation for any one analysis of 0.000015. Average in-run precision was 0.000015 for the standard and 0.000016 for each sample. A constant value of 0.00002 was subtracted from the data in this study for normalization with Hodell *et al.* (1991).

Carbon and oxygen isotopes were analyzed on a Finnigan-MAT 251 mass spectrometer in the Exploration and Production Technology Department of Texaco Oil Company in Houston, Texas. Data are reported as $\delta^{18}O$ and $\delta^{13}C$-values for CO_2 gas relative to PDB in the per mil standard notation $\delta^{18}O = [(Rsample\text{-}Rstandard)/Rstandard] \times 1000$ where $R = \delta^{18}O/\delta^{16}O$. One-third of the samples were run in duplicate, yielding an analytical precision of 0.1 per mil for O and 0.2 per mil for C. In-run reproducibility of results was determined relative to the NBS 20 Solenhofen Limestone standard, and yielded a precision of 0.1 per mil for O and 0.2 per mil for C. A fractionation factor of $\alpha T = [5.25\ (\pm 0.08) \times 10^2]T^{-2} + 1.003970\ (\pm 0.000034)$ for the common acid bath technique has been applied to calcite analyses (Swart *et al.*, 1991). No phosphoric acid fractionation correction has been applied to the dolomite analyses.

Carbon and oxygen isotope analyses of bladed calcite cements were performed by P.C. Smalley on the British Petroleum Laser Ablation Sampler for Stable Isotope Extraction (LASSIE) as described in Dickson *et al.* (1990, 1991). Based on analyses of natural and synthetic calcite standards at British Petroleum Sunbury Research Center, empirical corrections of −0.8‰ $\delta^{13}C$ and 1.2‰ $\delta^{18}O$ were applied to the laser data.

Cation abundances (Ca, Mg, Sr, Mn, Na and Fe) were determined on a Beckman Spectra Span VB DCP (Direct Current Argon Plasma) at Stony Brook with the assistance of S. McLennan. All samples were analyzed in triplicate, with analytical uncertainties and detection limits of each element reported in the data tables. Anions (Cl, F, and SO_4) were determined in collaboration with W. Staudt on a DIONEX 2010 Ion Chromatograph at Stony Brook (Staudt *et al.*, 1993). Analytical uncertainties and detection limits are reported in each data table.

In situ elemental abundances were obtained for Ca, Mg, Fe and Mn by electron probe microanalysis (EPMA) at Stony Brook using a Cameca Camebax electron probe operating at 15 kV and 0.015 μA (for Ca and Mg) or at 30 kV and 0.015 μA (for Fe and Mn). Ten replicate analyses were made for each spot and averaged results were used for all comparisons. Estimated precision is expressed as 95% confidence limits, which are calculated from replicate analyses. For Ca and Mg, these are typically 1.5%, or approximately 0.8 mol% MCO_3. For Mn and Fe, confidence limits are generally less than 4%.

The transition metals Zn and Cu were analyzed *in situ* on single crystals of dolomite within 30 to 50 m-thick doubly polished thin-sections mounted on Optisil pure-silica glass slides with synchrotron x-ray fluoresence microanalysis (SXRFMA) on beamline 26A at the National Synchrotron Light Source at Brookhaven National Laboratories in Upton, New York. The procedures for carbonate analysis used in this study were developed by Paquette (1992) and Ward (1993). Following the procedures in Kopp *et al.* (1990), concentrations of Zn and Cu were determined by extracting net peak areas with a peak-fitting program, which were then cross-checked with the standard called FLN housed at Stony Brook. The optimal detection limit for Zn and Cu was 2 ppm, with standard deviations of approximately 20%.

Groundwater samples were collected during October, 1989, from a total of 15 wells that penetrate the Curaçao Lava Formation in southeastern Curaçao. Samples were collected from the well head at each locality after allowing the pump to run for 15 minutes at high pressure. Water temperatures were measured during this period by inserting a thermometer into the well spout. The pH was directly measured by inserting a rugged combination electrode (Ag-AgCl reference) into the sample bottle. Total alkalinity was determined with a precision of 1% on 0.5 ml of water sample by titration with 0.025N HCl. Chlorine was determined with a precision of 1% by automated coulometric titration with Ag.

CHAPTER 2

SEDIMENTOLOGY AND DEPOSITIONAL ENVIRONMENTS

Introduction

In this chapter the sedimentology, stratigraphy, and depositional environments of the Seroe Domi Formation will be presented. Investigations of Seroe Domi Formation dolomitization on Bonaire (Deffeyes *et al.,* 1965; Murray, 1969; Bandoian and Murray, 1974; Sibley, 1980, 1982; Lucia and Major, in press) and phosphatization on Curaçao (Stienstra, 1983, 1991; Ten Have *et al.,* 1982) have referred to De Buisonjé (1974) for the basic geology of the Seroe Domi Formation. De Buisonjé described Older, Middle and Younger subdivisions, mapped the formation as a single unit, and interpreted it to be a Late Miocene to Pliocene and possibly Pleistocene forereef deposit. However, during preliminary mapping in this study the distinction and correlation of the Older, Middle and Younger lithologic subdivisions suggested by De Buisonjé (1974) were not consistently observed in the field. In addition, De Buisonjé (1974) provides no supporting measured sections, maps, or other field data to substantiate these subdivisions. Therefore, it has been necessary to establish the basic sedimentologic and chronostratigraphic framework in the present study.

Lithofacies of the Seroe Domi Formation

The Seroe Domi Formation is composed of three lithologically and stratigraphically distinct divisions, here in ascending chronologic order called Subunits 1, 2 and 3. The sedimentologic composition and distinctive properties of each subunit is presented in this section in order to construct a three-dimensional stratigraphic framework for the entire Seroe Domi Formation

SUBUNIT	1				2			3
LITHOFACIES	**1**	**2**	**3**	**4**	**5**	**6**	**7**	**8**
COMPOSITION	Coral-Algal Pack-Grainstone	Foraminifer Wacke-Packstone	Coral-Algal Grain-Boundstone	Algal-Coral Pack-Grainstone	Coral-Algal Grain-Boundstone	Siliciclastic Pack-Grainstone	Coral-Algal Grain-Boundstone	Siliciclastic Sandstones
LITHOLOGY								
Color	buff to white	brown to buff	gray to white	brown to buff	brown to buff	buff to brown	buff to brown	brown
Induration	variable, P to W	P to M	W	W	variable, P to W	variable, P to M	variable, P to W	variable, P, M
Grain Size	M, S, and G	M and S, some G	S and G	M, S, and G	M, S, and G	M, S, and G	M, S, and G	M, S, and G
Grain Sorting	P	M to W	P	P	P	P	P	variable, P, M
Grain Abrasion	U, R and A	U, R and A	U	U, R and A	U	U, R, and A	U, R, and A	R and A
% Clastics	variable, 0 to 30	variable, 0 to 30	< 5	variable, 0 to 30	variable, 10 to 30	variable, 25 to 60	variable, 0 to 20	80 to 100
% Dolomite	variable, 0 to 100	variable, 0 to 100	< 10	100	75 to 100	0	variable, 0 to 100	0
% Porosity	15 to 34	1 to 13	4 to 18	2 to 15	15 to 39	12 to 19	8 to 25	13 to 30
Mean Porosity	21%	9%	13%	10%	22%	12%	21%	21%
BEDDING								
Geometry	tabular to planar	tabular to planar	massive blocks	tabular to massive	tabular to massive	tabular to planar	tabular to planar	planar
Thickness	0.1 to 1 m	0.1 to 1 m	< 5 m x 8 m	0.1 to 3 m	0.1 to 2 m	0.1 to 1.5 m	0.1 to 1.5 m	0.1 to 1 m
Strike (degrees)	280 to 330 NW	280 to 330 NW		280 to 330 NW	5 to 10 NE	260 to 285 NW	260 to 285 NW	260 to 285 NW
Dip (degrees)	10 to 30 S	10 to 20 S		5 to 10 S	7 S to 10 N	10 to 23 S	10 to 23 S	0 to 5 S
STRUCTURES								
Primary	nongraded geopetals alloch. blocks	nongraded geopetals alloch. blocks	nongraded geopetals	nongraded geopetals	nongraded geopetals	nongraded	nongraded geopetals	nongraded truncated beds
Secondary	fractures	fractures	fractures	fractures	fractures		fractures	
BIOTA								
Assemblage	mixed B and P	mixed, primarily P	predominantly B	predominantly B	predominantly B	predominantly B	predominantly B	mostly B
Abundance	low	high	low	high	low	low	low	low
Diversity	high	low	high	low	high	low	high	low
Occurence	no G, all N	no G, all N	most G, some N	uncertain	some G, most N	no G, all N	some G, most N	no G, all N
DEPOSITION								
Environment	Forereef to Carbonate Slope	Forereef to Carbonate Slope	Forereef to Carbonate Slope	Forereef to Carbonate Slope	Shelf to Reef Front	Forereef to Reef Front	Reef Front	Fluvial to Shallow marine
Age	Middle Miocene	Middle Miocene	Middle Miocene	Middle Miocene	Late Mio-Pliocene	Late Mio-Pliocene	Late Pliocene	Early Pleistocene
DIAGNOSTIC CRITERIA	inclined bedding skeletal 87Sr/86Sr *M. imperatoris*	*P. glomerosa* high % micrite lack of corals	alloch. block *M. cavernosa* algae coatings	high % red algae lack of corals pervasive dolomite	concave bedding *M. limbata* pervasive dolomite	high % clastics inclined beds lack of corals	inclined bedding agaracid reefs skeletal 87Sr/86Sr	high % clastics horizontal beds *A. palmata*
TYPE SECTIONS	St. Michielsberg Seroe Blandan	St. Michielsberg Seroe Blandan	St, Michielsberg Seroe Blandan	Seru Boka Duivelsklip	Seru Mansinga	Seru Spreit	St. Annabaai Seru Spreit	St. Annabaai

Abbreviations

Induration: P = poor; M = moderate; W = well

Grain Size: M = mud; S = sand; G = gravel

Grain Sorting: P = poor; M = moderate; W = well

Grain Abrasion: U = unabraded; R = rounded; A = angular

Assemblage: B = benthic; P = pelagic

Occurence: G = in growth position; N = not in growth position

TABLE 2.1. Sedimentology, paleontology, depositional interpretations, diagnostic criteria, and type section localities for the Seroe Domi Formation on Curaçao.

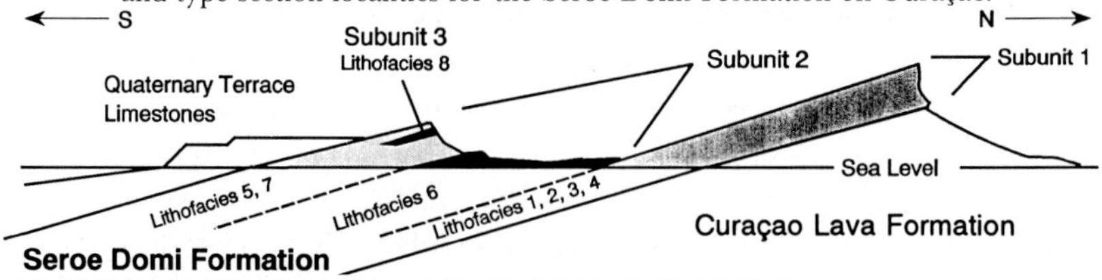

Subunit 1: Middle Miocene Interbedded Coralgal Grainstones, Foraminifer Wackestones, and Allochthonous Blocks
Lithofacies 1, 2, 3, and 4
Oyster 87Sr/86Sr = 0.708781 - 0.708832
Pelagic Foraminifer *Praeorbulina glomerosa circularis* Blow
Benthic Foraminifer *Melonis pompiloides* Fitchell and Moll
Lower Subunit 2: Late Miocene to Pliocene Siliciclastic Packstones and Grainstones
Lithofacies 6
Oyster 87Sr/86Sr = 0.709008
Upper Subunit 2: Pliocene to Early Pleistocene Coralgal Grainstones and Boundstones
Lithofacies 5 and 7
Brachiopod and Oyster 87Sr/86Sr = 0.709045 - 0.709063
Scleractinian Biostratigraphic Range Overlap
Subunit 3: Early Pleistocene Siliciclastic Sandstones and Grainstones
Lithofacies 8
Deposited in Eroded Upper Subunit 2 Beds
Scleractinian *Acropora palmata*

FIGURE 2.1. Geologic cross-section summarizing the Curaçao Lava Formation, Seroe Domi Formation, and Quaternary Limestone Terraces succession on Curaçao. Cross-section is oriented perpendicular to the leeward coast of Curaçao at St. Michielsberg. Listed below the cross-section is a summary of the sedimentologic and chronostratigraphic evidence observed in Subunits 1, 2, and 3 of the Seroe Domi Formation.

PALEONTOLOGIC ASSEMBLAGES	SUBUNIT 1			
	LITHOFACIES 1	**LITHOFACIES 2**	**LITHOFACIES 3**	**LITHOFACIES 4**
SCLERACTINIANS	*Agaracia cf. lamarcki* *Montastrea imperatoris* *Porites cf. astreoides* *Porites cf. porites* *Siderastrea sp.*	*Agaracia cf. lamarcki* *Montastrea imperatoris* *Porites cf. astreoides* *Porites cf. porites* *Siderastrea sp.*	*Agaracia agaricites* *Montastrea cavernosa* *Montastrea limbata* *Siderastrea siderea*	*Siderastrea sidera* *Montastrea limbata*
FORAMINIFERS	*Amphistegina sp.* *Homotrema sp.*	*Globorotalia foshi peripheroronda* *Globorotalia mayeri* *Praeorbulina glomerosa circularis*	*Amphistigina sp.* *Homotrema sp.*	*Amphistegina sp.*
BIVALVES	*Ostrea frons* *Lithophagus sp.*			
BRACHIOPODS				

PALEONTOLOGIC ASSEMBLAGES	SUBUNIT 2			SUBUNIT 3
	LITHOFACIES 5	**LITHOFACIES 6**	**LITHOFACIES 7**	**LITHOFACIES 8**
SCLERACTINIANS	*Acropora panamaensis* Agaraciidae fragments *Dichocoenia stokesii* *Leptoseris sp.* *Montastrea limbata* *Montastrea trinitatis* *Porites cf. baracoaensis* *Porites porites* *Psammocora sp.* *Siderastrea siderea* *Stylophora canalis* *Stylophora imperatoris* *Stylophora macdonaldi* *Stylophora sp?*	*Montastrea limbata* *Porites porites* *Siderastrea sp.*	*Acropora panamaensis* *Agaracia lamarcki* *Agaraciidae fragments* *Colpophyllia sp.* *Dichocoenia stokesii* *Diploria clivosa* *Leptoseris cf. walli* *Leptoseris sp.* *Monastrea annularis* *Montastrea lombata* *Stephanocoenia spongiformis*	*Acropora cervicornis* *Acropora palmata*
FORAMINIFERS	*Amphisitigina sp.*		*Amphistigina sp.* *Homotrema sp.*	*Amphistigina sp.*
BIVALVES	*Ostrea frons* *Lithophagus sp.*	*Ostrea frons*	*Ostrea frons* *Lithophagus sp.*	
BRACHIOPODS			*Thecidellina barretti*	

TABLE 2.2. Paleontological assemblages observed in the present study of the Seroe Domi Formation on Curaçao.

outcrop belt on Curaçao. The Seroe Domi Formation overlies basalt and siliciclastic lithologies comprising the Curaçao basement, and is unconformably onlapped by Quaternary Limestone Terraces (Fig. 2.1). A composite measured section of the Seroe Domi is presented in Fig. 2.2. Descriptions of the sedimentology, bedding characteristics and faunal assemblages of the eight lithofacies comprising these subunits are presented below and summarized in Tables 2.1 and 2.2, and Fig. 2.3.

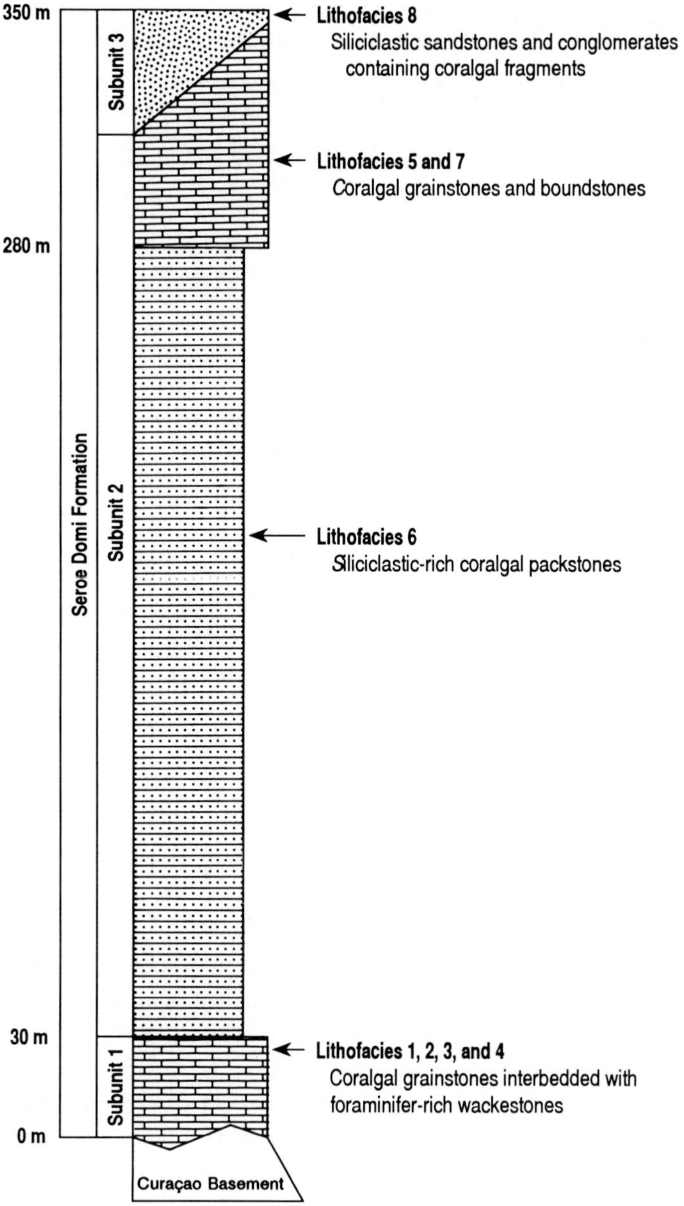

FIGURE 2.2. Composite stratigraphic column of the Seroe Domi Formation on Curaçao.

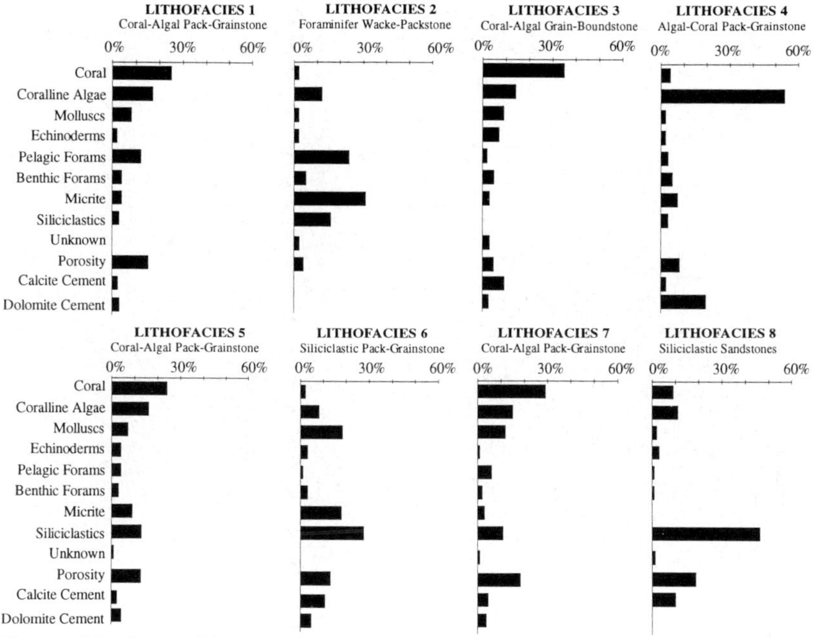

FIGURE 2.3. Compositional histograms depicting the relative proportions of grains, cements and porosity comprising the Seroe Domi Formation. Histograms are based on thin-section point count analyses (n = 500) averaged from 3 separate samples of each lithofacies.

Subunit 1 Lithofacies 1, 2, 3, and 4

Subunit 1, the 30 to 100 m-thick basal portion of the Seroe Domi Formation, is composed of a sequence of grainstone and packstone carbonate conglomerates interbedded with wackestones and allochthonous blocks, herein called Lithofacies 1, 2, 3, and 4.

Lithofacies 1 is composed of buff to white conglomeratic coralgal packstones and grainstones distinguished by their abundance of displaced coral heads of *Montastrea imperatoris* and *Porites cf. astreoides* within coarse pelagic foraminifer-rich skeletal sands (Tables 2.1 and 2.2; Fig. 2.3). The type section for Lithofacies 1 crops out at Seru Blandan (Location 26 in Fig. 1.8). Sediments occur as 0.1 to 1 m-thick non-graded tabular beds with seaward dips of 10°

FIGURE 2.4. Photographs of Subunit 1 Lithofacies 1, 2, and 3 type-section exposures at Seru Blandan and St. Michielsberg. A. Southeastern face of Seru Blandan. B. Northern view looking directly at the apex of Seru Blandan. The white boulder at the base of the

flatiron apex is the Lithofacies 3 allochthonous block. C. Southeastern face of St. Michielsberg. D. Northern view looking directly at the apex of St. Michielsberg. The white boulder at the base of the flatiron apex is the Lithofacies 3 allochthonous block.

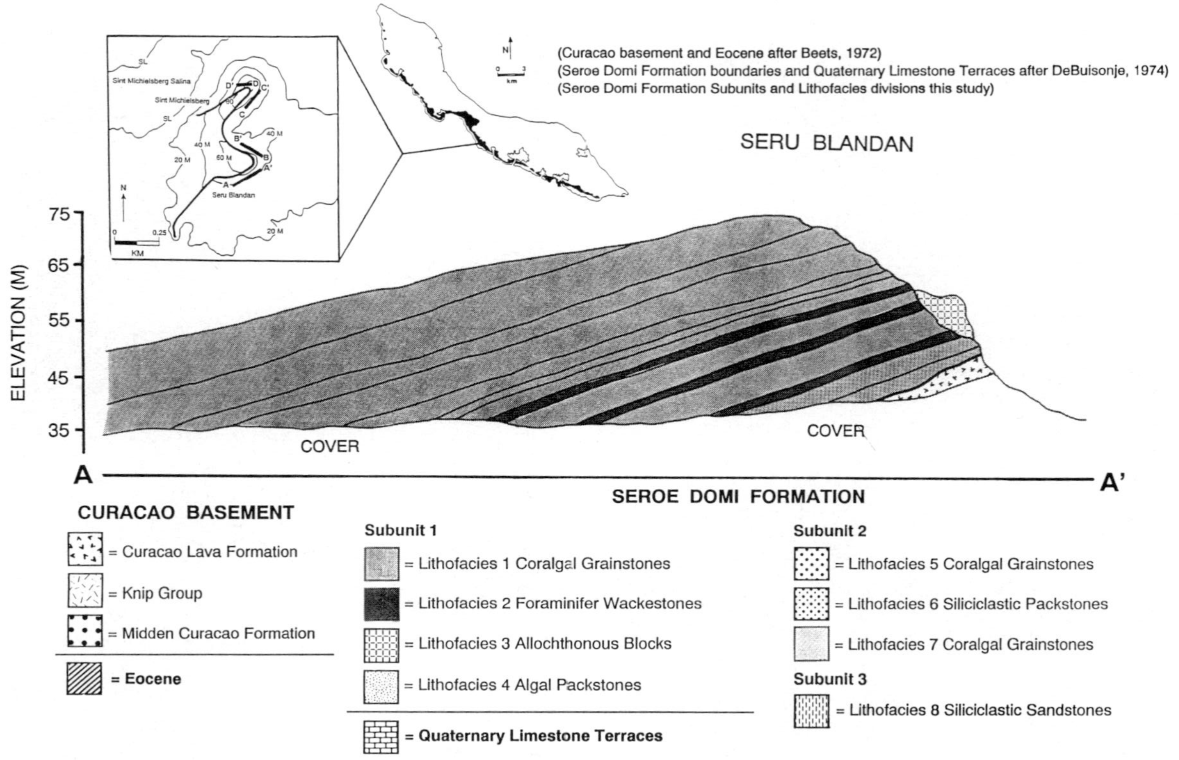

FIGURE 2.5. Line drawing of the Subunit 1 Lithofacies 1, 2, and 3 type-section outcrop at Seru Blandan. This illustration is traced from the field photograph in Figure 2.4A.

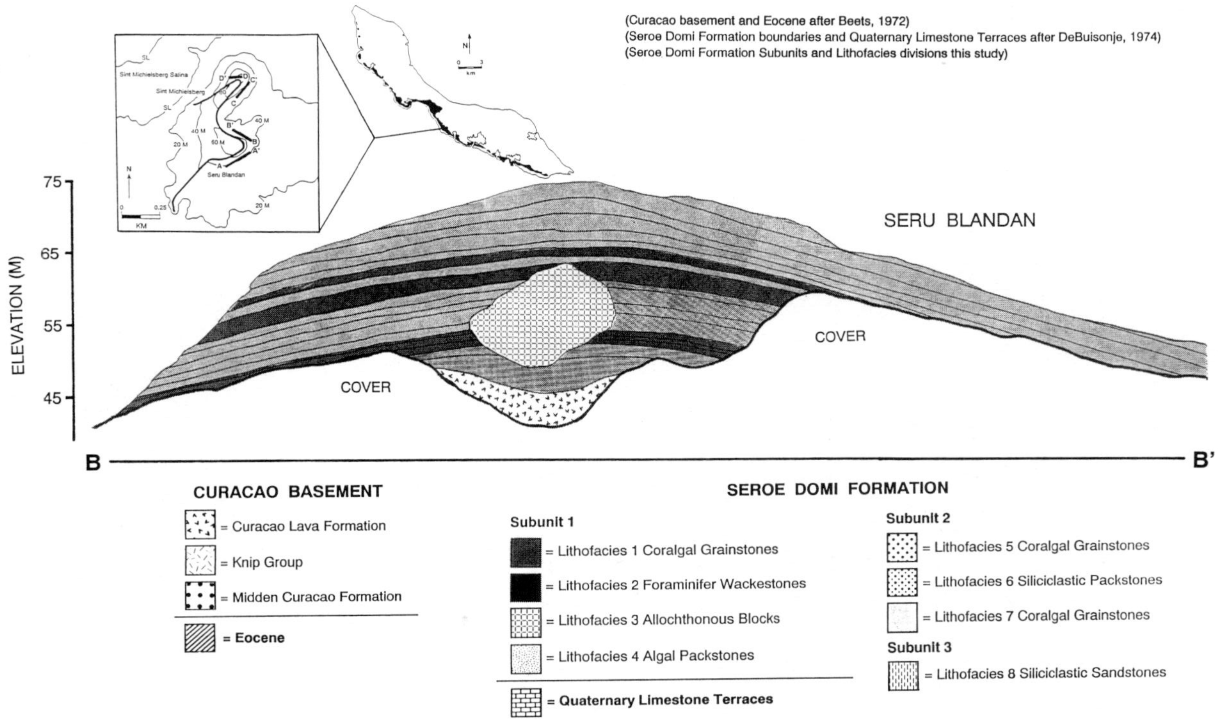

FIGURE 2.6. Line drawing of the Subunit 1 Lithofacies 1, 2, and 3 type-section outcrop at Seru Blandan. This illustration is traced from the field photograph in Figure 2.4B.

to 30° S and strikes of 280° to 330° NW (Figs. 2.4, 2.5 and 2.6). The basement contact is irregular with up to 3 meters of relief, and densely fractured with Lithofacies 1 sands and muds filling the fractures. Interbedded wackestones (Lithofacies 2) and allochthonous blocks (Lithofacies 3) are present in the lower 15 m of some Lithofacies 1 sections such as Seru Blandan and Sint Michielsberg (Location 25 in Fig. 1.8). However, they are absent in many other localities such as Seru Sami and Veerisberg (Locations 23 and 29 in Fig. 1.8). Lithofacies 1 skeletal sands are composed of fragments of corals, coralline algae, pelagic foraminifers, molluscs, benthic foraminifers, echinoderms, intraclasts, and basalt-derived siliciclastic sands and clasts. All of these components exhibit variable degrees of dissolution and abrasion. Corals, bivalves and echinoderms are completely dissolved, while foraminifers, oysters and red algae are often petrographically unaltered. Several coral heads contain elongate *Lithophagus sp.* borings that are partially to completely filled with sand and mud, forming geopetals that are aligned to within ±5° of present-day horizontal (De Buisonjé, 1974).

Lithofacies 2 is composed of light brown to buff foraminifer wackestones and packstones that occur as 0.1 to 1 m-thick tabular beds conformably interbedded with Lithofacies 1 (Tables 2.1 and 2.2; Fig. 2.3). The Lithofacies 2 type section is also at Seru Blandan. Interbedded Lithofacies 2 horizons form sharp upper and lower contacts with the surrounding Lithofacies 1 packstones and grainstones. Lithofacies 2 sands are primarily composed of pelagic and benthic foraminifers, but also contain fragments of coralline algae, corals, molluscs and echinoderms supported in gray muds. Point-counts of Lithofacies 2 thin sections suggest a 3:1 ratio of pelagic to benthic foraminifers (Fig. 2.3). Foraminifers tests appear unbroken, while red algae, coral and other skeletal material occur as irregular fragmented skeletal material. Angular siliciclastic gravels and boulders also occur locally within Lithofacies 2 wackestones. Lithofacies 2 is also locally bioturbated, containing trace fossils similar in form to *Zoophycos sp.* and *Chondrites sp.* (Seilacher, 1978; Ekdale *et al.*, 1984), which are best preserved at Fort Nassau 2 m above the basement contact (Location 34 in Fig. 1.8). The Lithofacies 2 foraminifer fauna include the pelagic species *Praeorbulina glomerosa circularis* (Blow), *Globorotalia foshi peripheroronda*, and *Globorotalia mayeri* (Fig. 2.7) and the benthic species *Laticarrinina pauperata* (Fitchel and Moll) and *Melonis pompiloides* (Fitchel and Moll; Fig. 2.8).

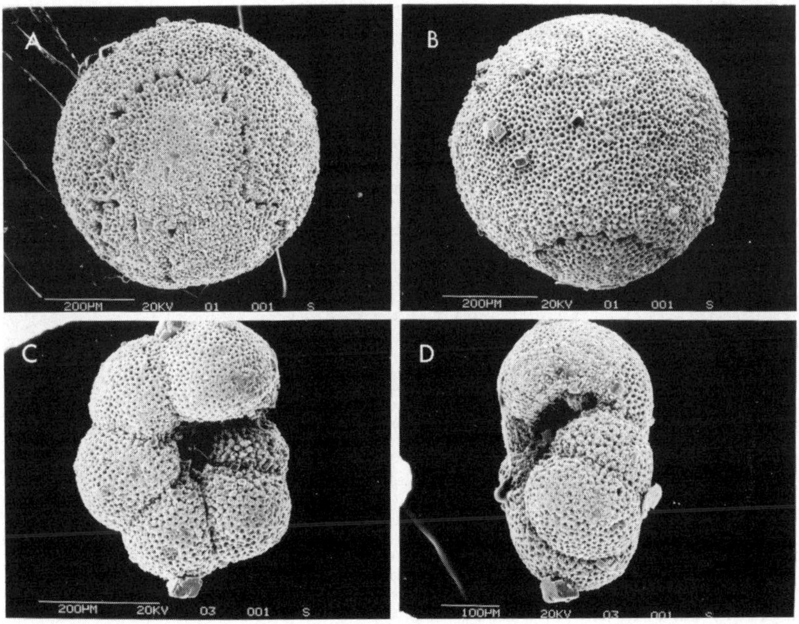

FIGURE 2.7. Scanning electron micrograph images of pelagic foraminifers in Subunit 1 Lithofacies 2 wackestones at Seru Blandan. A. *Praeorbulina glomerosa circularis* (Blow). B. *Praeorbulina glomerosa circularis* (Blow). C. *Globorotalia mayeri* . D. *Globorotalia mayeri*.

Lithofacies 3 occurs exclusively as 0.3 to 8 m-diameter allochthonous blocks within Lithofacies 1 and 2 deposits, and consists of light gray coralgal grainstones with large 0.5 to 0.75 m diameter heads of *Montastrea cavernosa* and other massive corals (Tables 2.1 and 2.2; Fig. 2.3). The type section for Lithofacies 3 is also at Seru Blandan. Lithofacies 3 sands are composed of fragments of corals, coralline algae, molluscs, echinoderms, benthic foraminifers, micrite and pelagic foraminifers (Fig. 2.3). Many coral heads are encrusted with 0.5 to 5 mm-thick coatings of crustose coralline algae, while the outer surfaces of the blocks have patchy 0.5 to 2 mm-thick crusts of coralline algae. Lithofacies 2 wackestones underlying the St. Michielsberg block exhibit distorted bedding that conforms to the shape of the lower surface of the block.

Lithofacies 4 is composed of buff to white pervasively dolomitized coralline algae packstones (Tables 2.1 and 2.2; Fig. 2.3). Lithofacies 4 is distinguished

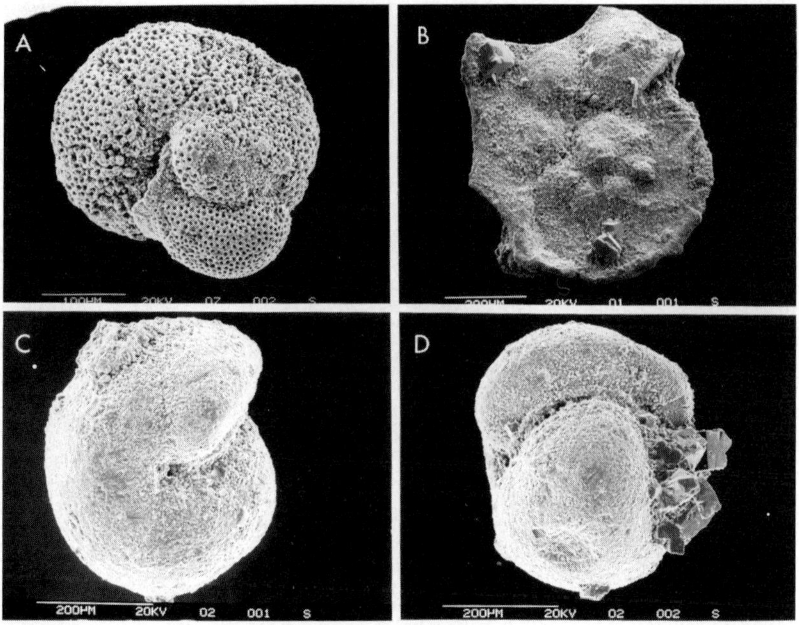

FIGURE 2.8. Scanning electron micrograph images of pelagic and benthic foraminifers in Subunit 1 Lithofacies 2 wackestones at Seru Blandan. A. The pelagic species *Globorotalia foshi peripheroronda*. B. The benthic species *Laticarrinina pauperata* (Fichtel and Moll). C. The benthic species *Melonis pompiloides* (Fichtel and Moll). D. *Melonis pompiloides* (Fichtel and Moll).

from Lithofacies 1 by the abundance of red algae, relative scarcity of corals, and restricted geographic distribution. Although Lithofacies 4 has undergone fabric-destructive dolomitization, the red algae are well preserved and inclusion ghosts in the sucrosic dolomites have shapes resembling skeletal fragments. Lithofacies 4 crops out as geographically and stratigraphically isolated massively dolomitized mountains at Seru Magasina, Tafelberg, Ser'i Boka, and Kabrietenberg, with the type-section at Duivelsklip (Locations 38 to 42, and 44 in Fig. 1.8). Lithofacies 4 lithologies are heavily fractured, with cracks reaching 0.5 m in width and 20 m in length that are filled with laminated blocky calcite cements. However, the fractures do not correlate with fractures in the basement lithologies and the overlying lithologies have been eroded. Therefore, it is not possible to determine the relative timing of these fracturing events with respect to the younger Seroe Domi Formation deposits.

Subunit 2 Lithofacies 5, 6, and 7

The Subunit 2 division is composed of Lithofacies 5, 6 and 7 and comprises the upper 250 to 300 m of the Seroe Domi Formation.

Lithofacies 5 is composed of pervasively dolomitized light brown coralgal packstones, grainstones and boundstones that crop out exclusively at Seru Mansinga (Location 38 in Fig. 1.8; Tables 2.1 and 2.2; Fig. 2.3). Lithofacies 5 is distinct from Lithofacies 1 in having a unique coral assemblage, and distorted concave bedding that dips both seaward and landward (7° S to 10° N) and strikes 5° to 10° NE (Figs. 2.9 and 2.10). And unlike the fabric-destructive dolomitization of Lithofacies 4, Lithofacies 5 exhibits exceptional fabric-replacive (mimetic) pervasive dolomitization. The 20 to 30 m-thick lower half of the Seru Mansinga section is dominated by 0.1 to 0.5 m-thick planar packstone horizons of the corals *Porites porites*, *Montastrea trinitatis*, *Stylophora sp.* and *Leptoseris sp.*. These are interbedded with coarse to fine sands composed of fragments of corals, coralline algae, siliciclastics, molluscs, echinoderms, crustaceans, benthic and pelagic foraminifers, and micrite. The 20 to 25 m-thick upper half of the section is composed of boundstones containing *in situ* branching and hemispherical forms of the coral *Montastrea limbata*. Gray to pink dolomitized internal sediments are also common throughout both the upper and lower parts of the section.

Lithofacies 6 is composed of buff to light brown packstones and grainstones (Tables 2.1 and 2.2; Fig. 2.3). Lithofacies 6 is distinguished from Lithofacies 1 by their high content of siliciclastic sands, decreased coral abundance, and significant changes in strike. Lithofacies 6 exposures at Seru Spreit conformably overlie Subunit 1 Lithofacies 1, and are preferentially eroded to form soil and brush-covered lows between ridges composed of Lithofacies 1 and Lithofacies 7 (Location 24 in Fig. 1.8; Fig. 2.11) . The 0.1 to 0.5 m-thick planar to tabular bedding of Lithofacies 6 dips 10° to 25° S and strikes 265° to 285° NW. Lithofacies 6 sands are composed of up to 50 percent siliciclastics, whole and fragmented bivalves, calcareous algae, echinoderms, corals, and benthic foraminifers. However, they contain few pelagic foraminifers relative to Subunit 1 lithologies. With the exception of a few oysters and benthic foraminifers, Lithofacies 6 skeletal sands have been extensively leached, and the voids and molds almost completely filled with blocky calcite ce-

FIGURE 2.9. Photographs of the Subunit 2 Lithofacies 5 dolomitized coralgal packstones and grainstones type-section exposure at Seru Mansinga. Note the distinct differences in bedding between the planar coralgal packstone lithologies (p) in the lower half of the section and the massive coralgal boundstones (m) in the upper half of the section in each of the photographs. A. Eastern face of Seru Mansinga looking westward across Caracasbaai. B. Close-up view of region marked "a" in photograph A, illustrating the concave-upward

bedding in the northern area of the outcrop and seaward dips to the south of exposure. C. Close-up view of region marked "b" in photograph A, showing convex-upward bedding in the central part of the outcrop. D. Close-up view of region marked "c" in photograph A, illustrating seaward dips of both the lower coralagal packstones (p) and the upper coralgal boundstones (m) at the southern end of the outcrop.

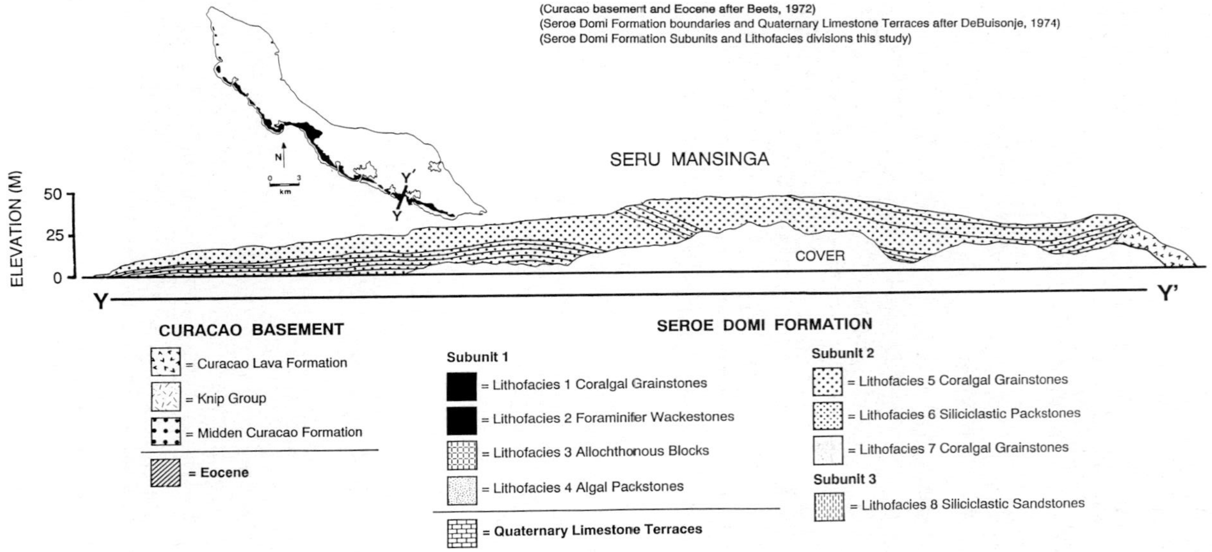

FIGURE 2.10. Line drawing of the Subunit 1 Lithofacies 5 type-section outcrop at Seru Mansinga. This illustration is traced from the field photograph in Figure 2.9A.

FIGURE 2.11. Photographs of Subunit 2 Lithofacies 6 siliciclastic-rich coralgal packstones to grainstones type-section exposure at Seru Spreit. A. Oblique view looking toward the southwest from St. Michielsberg at the Subunit 1 Lithofacies 1 (L1) exposures at Seru Spreit that are conformably overlain by Subunit 2 Lithofacies 6 (L6) and Lithofacies 7 (L7). Lithofacies 6 lithologies (L6) are preferentially eroded, forming a low between ridges composed of Lithofacies 1 (L1) and Lithofacies 7 (L7) limestones. B. The eroded and partially covered contact between Subunit 1 Lithofacies 1 (L1) and Subunit 2 Lithofacies 6 (L6) on the southern down-dip flank of the Seru Spreit flatiron. This is the area directly beneath the "L6" label shown in photograph A.

ments. Slender sinuous trace fossils 2 cm in diameter and up to 1 m in length occur in packstones near the base of the Lithofacies 6 section at Seroe Spreit.

Lithofacies 7 is composed of buff to white conglomeratic coralgal packstones, grainstones, and boundstones, which are well-exposed on the west bank of the Sint Annabaai channel (Location 33 in Fig. 1.8). Lithofacies 7 grainstones are similar in fabric and composition to Lithofacies 1 grainstones. Lithofacies 7 is distinguished primarily by having relatively low abundances of pelagic foraminifers, a different coral assemblage (Tables 2.1 and 2.2; Fig. 2.3), and the lack of interbedding with Lithofacies 2 wackestones. Lithofacies 7 packstones and grainstones also contain variable amounts of siliciclastic debris, abundant fragments of *Acropora panamaensis*, *Agaracia lamarcki*, *Diploria clivosa*, *Dichocoenia stokesii*, and *Montastrea lombata* in coarse skeletal sands. Lithofacies 7 deposits form 0.5 to 1 m-thick tabular beds striking 265

FIGURE 2.12. Photographs of the Subunit 2 Lithofacies 7 coralgal grainstones and Subunit 3 Lithofacies 8 siliciclastic sandstones type-section outcrop at St. Annabaai. A. Western bank of the St. Annabaai channel, illustrating stratigraphic relationships of Subunit 2 Lithofacies 6 siliciclastic-rich packstones (L6), Lithofacies 7 coralgal grainstones (L7), Subunit 3 Lithofacies 8 siliciclastic sandstones and conglomerates (L8), and an overlying deposit of Quaternary Limestones Terraces (QLT). B. Close-up view of region marked "a" in photograph A, illustrating conglomeratic debris (d) composed of sands, cobbles and boulders deposited in an irregular erosional void within inclined Lithofacies 7 coralgal grainstones (L7). C. Close-up view of region marked "b" in photograph A. Lithofacies

8 siliciclastic sandstones and conglomerates occur with two seperate erosional troughs within Lithofacies 7 coralgal grainstones. A smaller stratiform trough filled with inclined Lithofacies 8 sandstones (L81) is truncated by a wide erosional void containing horizontally bedded Lithofacies 8 sandstones and conglomerates (L82). D. Conformable gradational contact between Subunit 2 Lithofacies 6 siliciclastic-rich packstones (L6) and Lithofacies 7 coralgal grainstones (L7). Arrow points to vertical paleosol "pipes" within Lithofacies 6 filled with red lateritic soils and Pleistocene *Acropora palmata* coral debris, suggesting that they were not formed until the post-Lithofacies 7 deposition of the Quaternary Limestone Terraces.

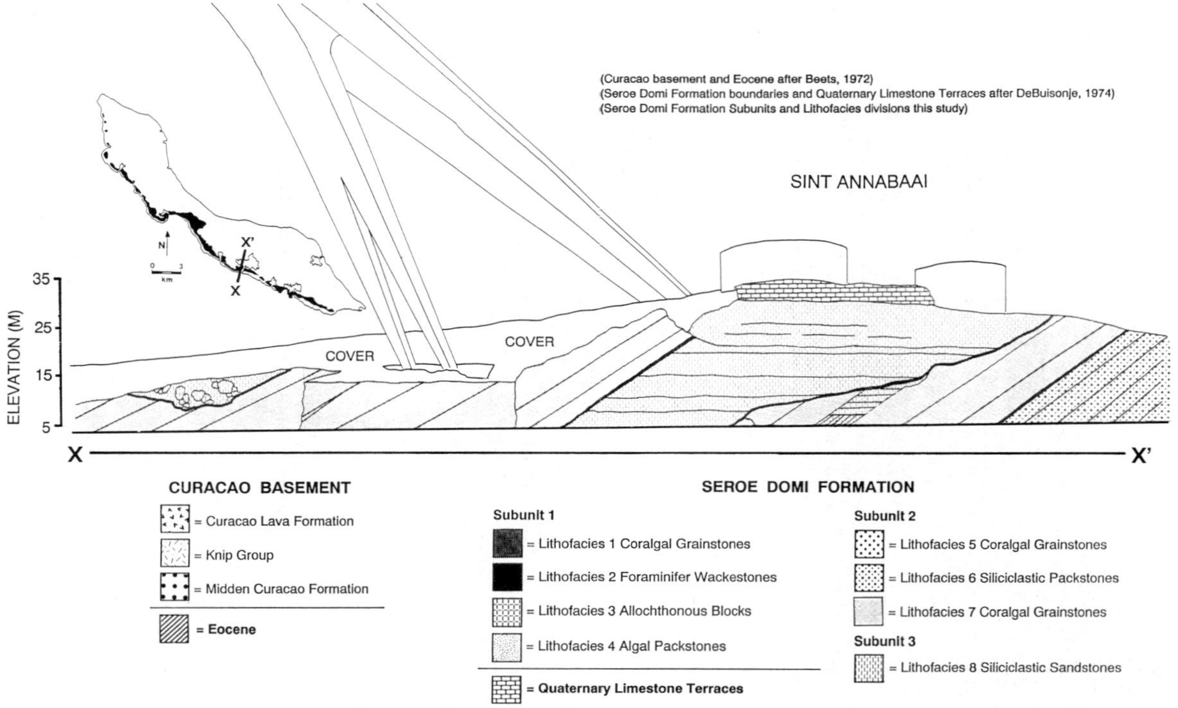

FIGURE 2.13. Line drawing of the Subunit 2 Lithofacies 7 and Subunit 3 Lithofacies 8 type-section outcrop at St. Annabaai. This illustration is traced from the field photograph in Figure 2.12A.

to 280° NW and dipping 10 to 23° S (Figs. 2.12 and 2.13). The boundstones are 0.5 to 1 m-thick wedge-shaped beds composed of *in situ Agaracia lamarcki* and *Leptoseris cf. walli* fronds encrusted with *Thecidellina barretti* brachiopods in densities of up to 1000/m². Corals, bivalves and echinoderms have been extensively leached, while oysters and brachiopods have resisted dissolution. Lithofacies 7 grainstones have been extensively eroded and perhaps locally karsted, forming both stratiform and bed cross-cutting chutes that have been filled with Lithofacies 8 sandstones and conglomerates (Figs. 2.12 and 2.13).

Subunit 3 Lithofacies 8

Subunit 3 is composed of siliciclastic sandstones and conglomerates, herein called *Lithofacies 8*, that have been deposited in laterally confined, selectively eroded stratiform and bed cross-cutting cavities within Lithofacies 7 deposits. The type section for Lithofacies 8 is the same as that of Lithofacies 7 on the west bank of the Sint Annabaai channel (Location 33 in Fig. 1.8). Lithofacies 8 exhibits two distinct depositional sequences at St. Annabaai, that include: 1) an older sequence filling a stratiform erosional void wihin Lithofacies 7, consisting of a lower 1.5 m-thick horizon of 30° inclined planar beds truncated by an overlying 3 m-thick horizontally bedded deposit; and 2) a younger horizontally bedded sequence 9 m-thick that truncates the older unit and occurs within a bed cross-cutting void in Lithofacies 7 (Figs. 2.12 and 2.13). Lithofacies 8 is composed of siliciclastic sands, irregularly shaped angular to rounded clasts of Curaçao Lava Formation diabase 0.05 to 0.75 m in diameter, as well as large abraded cobbles of *Acropora palmata* and *Acropora cervicornis* (Tables 2.1 and 2.2; Fig. 2.3). These coral and diabase cobbles are sometimes coated with 0.5 to 5 mm-thick layers of encrusting coralline algae. Unlike Subunits 1 and 2 lithologies, Subunit 3 Lithofacies 8 lacks mud and consists of coarse sands and cobbles of siliciclastics and dissolved fragments of coralline algae, molluscs, corals and echinoderms filled with isopachous rims of blocky calcite cements.

Tafelberg Phosphatization

The Tafelberg mountain on Curaçao is the largest and topographically highest exposure of the Seroe Domi Formation (Location 41 in Fig. 1.8). The Tafelberg is an approximately 180 m-thick section of limestones that have experienced pervasive fabric-destructive phophatization, calcitization and dolomitization. The complex diagenetic sequence of the Tafelberg is petrographically and geochemically analyzed by Ten Have *et al.* (1982) and Stienstra (1983; 1991). Both workers interpret the phosphatization to be the result of meteoric alteration, with the phosphate having been derived from bird, bat, and marine guano deposits. De Buisonjé (1974) describes corals and bivalves from this locality, interprets the Tafelberg outcrop to be part of his Middle Seroe Domi Formation. However, extensive mining has removed or covered a significant portion of the Tafelberg, while pervasive fabric-destructive recrystallization has destroyed most of the original depositional texture. Therefore, during the present study no sedimentologic evidence was found to correlate the original coralgal limestones of the Tafelberg with either Subunit 1, 2, or 3. As a result, interpretation of the Tafelberg lithologies in this study will be limited to considering them to be Seroe Domi Formation deposits of unknown age and depositional affinity.

Quaternary Limestone Terraces

The Quaternary Limestone Terraces comprise a sequence over 60 m-thick of buff to white coralgal packstones, grainstones, and boundstones that unconformably onlap both the basement and the Seroe Domi Formation (Fig. 2.14). The Quaternary Terrace Limestones contain a high-diversity coral assemblage dominated by *Montastrea annularis*, *Acropora palmata*, *Siderastrea sidera* and *Porites porites*, as well as sands composed of fragments of coralline algae, molluscs, benthic foraminifers, echinoderms, pelagic foraminifers, and siliciclastic sands and gravels. De Buisonjé (1974) subdivided the Quaternary Terrace Limestones into three facies, which in a landward to seaward progression include: 1) the *Siderastrea* Zone (lagoonal); 2) the *Montastrea annularis* Zone (lagoonal); and 3) the *Acropora palmata* Zone (fringing reef).

The Quaternary Limestone Terraces were deposited on both the windward and the leeward sides of the island in a series of five horizontal stair-stepped benches or terraces (Fig. 2.14). These include the Lower, Middle I, Middle II, Higher, and Highest Terraces, each of which represents the multiple overprinting of depositional and erosional events (Alexander, 1961; De Buisonjé, 1974; Bandoian and Murray, 1974; Herweijer and Focke, 1978). The terraces that are lowermost in elevation are best preserved, most aerially

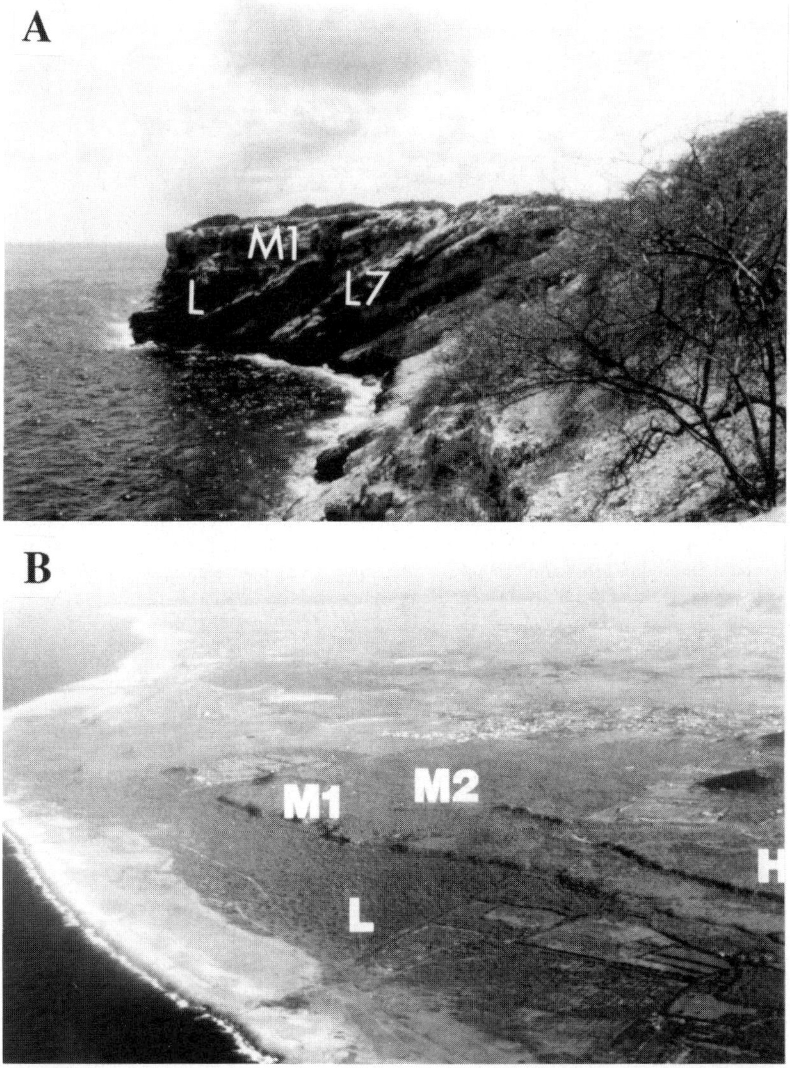

FIGURE 2.14. Photographs of the Quaternary Limestone Terraces. A. Lower (L) and Middle I (M1) Quaternary Limestone Terraces directly onlapping Subunit 2 Lithofacies 7 coralgal grainstone (L7) deposits of the Seroe Domi Formation at Blauwbaai. B. Oblique aerial photograph of the southeastern half of Curaçao, illustrating the stratigraphic and spatial relationships of the Lower (L), Middle I (MI), Middle II (MII), and Higher (H) Quaternary Limestone Terrace sequence. St. Joris Baai and Tafelberg are visible in the background.

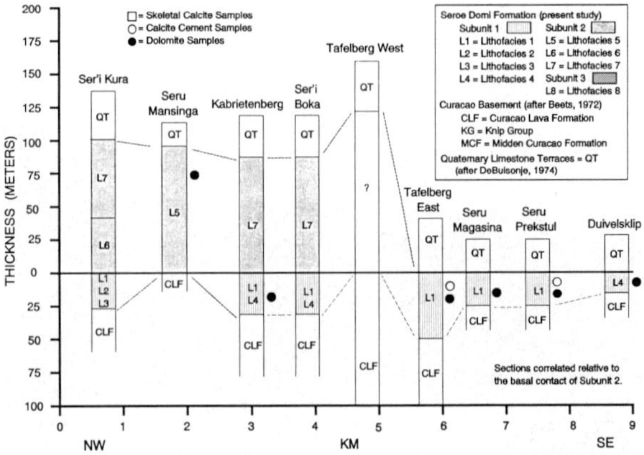

FIGURE 2.15. Lithostratigraphic diagram of the Seroe Domi Formation from Ser'i Kura to Duivelsklip. The diagram correlates the distribution of Subunits 1, 2, and 3 within the Seroe Domi Formation (present study), Cretaceous and Tertiary lithologies comprising the basement (after Beets, 1972), and the Quaternary Limestone Terraces (after De Buisonjé, 1974). The stratigraphic elevations of shell material, calcite cements and dolomite cements chosen for geochemical analysis are depicted with open squares, open circles, and darkened circles respectively.

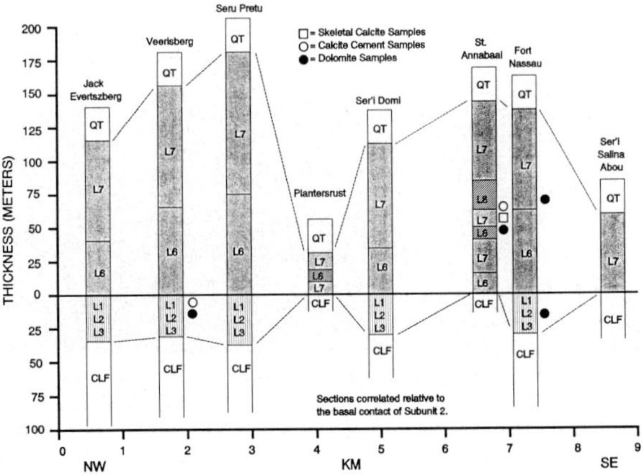

FIGURE 2.16. Lithostratigraphic diagram of the Seroe Domi Formation from Jack Evertzberg to Ser'i Salina Abou. The description and lithologic legend are in Figure 2.15.

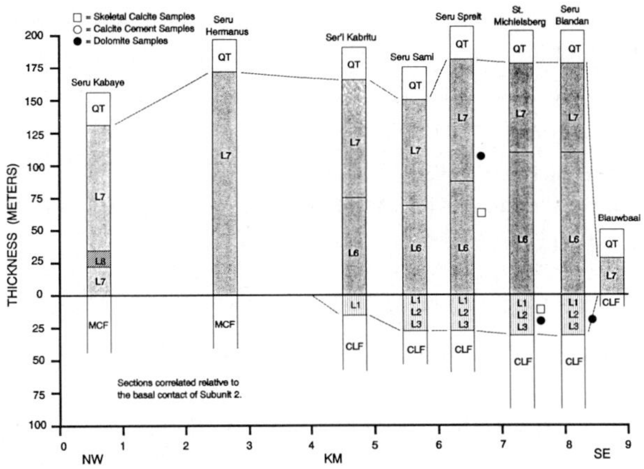

FIGURE 2.17. Lithostratigraphic diagram for the Seroe Domi Formation from Seru Kabaye to Blauwbaai. The description and lithologic legend are in Figure 2.15.

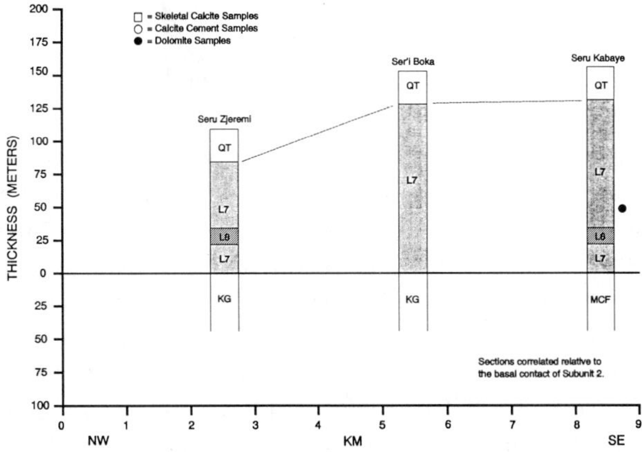

FIGURE 2.18. Lithostratigraphic diagram for the Seroe Domi Formation from Seru Zjeremi to Seru Kabaye. The description and lithologic legend are in Figure 2.15.

extensive, and have been interpreted to be the youngest deposits (De Buisonjé, 1974). Conversely, the higher elevation terraces, which are more poorly preserved and often occur as aerially restricted erosional remnants, have been interpreted as the older terraces (De Buisonjé, 1974; Herweijer and Focke, 1978). The topographically highest terrace deposit to directly onlap the Seroe Domi Formation is the Middle I Terrace, which is estimated to be at least 470,000 years old (Herweijer and Focke, 1978).

Lithostratigraphic Distributions and Correlations

Lithofacies outcrop maps at a scale of 1:25,000 and 1:15,000 are presented in Fouke (1993) for the entire Seroe Domi Formation outcrop belt from Seru Belefa to Duivelsklip (Localities 1 to 44 in Fig. 1.8). Measured sections in a series of lithostratigraphic diagrams correlated with respect to the contact between Subunits 1 and 2 are presented in Figs. 2.15, 2.16, 2.17, and 2.18. Areas between each Seroe Domi Formation flatiron mountain on these diagrams represent streams, gullies and other topographic lows in which Curaçao basement lithologies are exposed. Lithologic equivalency within the Seroe Domi Formation is based on the sedimentologic compositions and faunal assemblages of each lithofacies, lateral projections of bedding planes, and the relative elevation of each lithologic horizon with respect to the basement contact and present-day sea level.

Subunit 1 is aerially restricted to the southeastern half of Curaçao from Ser'i Kabritu to Duivelsklip (Fig. 2.19). Within this region, Lithofacies 1, 2 and 3 are found to the northwest of Seru Mansinga and at Seru Magasina and Seru Prekstul. Lithofacies 4 is interspersed with Lithofacies 1 in the southeast, and is restricted to Kabrietenberg, Ser'i Boka, Tafelberg and Duivelsklip. Lithofacies 2 is confined to the flatirons northwest of Ser'i Salina Abou, and Lithofacies 3 allochthonous blocks are present exclusively at Sint Michielsberg, Seru Blandan, and Ser'i Domi (Fig. 2.20). Subunit 2 Lithofacies 5, which occurs only at Seru Mansinga, is stratigraphically isolated from other Seroe Domi Formation deposits and onlapped by Quaternary Limestone Terraces (Fig. 2.21). Lithofacies 4 is also stratigraphically isolated, with the exception of the outcrops at Kabrietenberg (Fig. 2.21), where they are overlain by Lithofacies 7 grainstones as well as Quaternary Limestone Terraces.

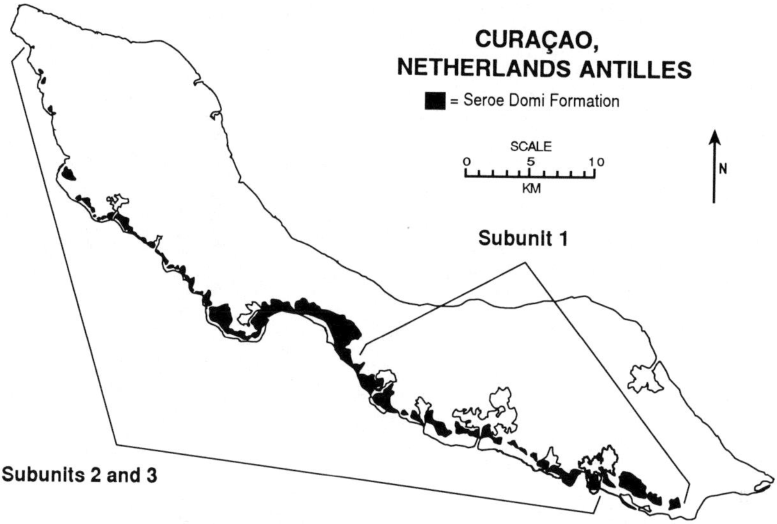

FIGURE 2.19. Geographic distributions of Seroe Domi Formation Subunits 1, 2, and 3 on Curaçao.

In contrast, Subunit 2 Lithofacies 6 and 7 crop out over the northwestern three-fourths of the island from Seru Belefa to Ser'i Kura, and from Seru Mansinga and Kabrietenberg (Fig. 2.19). These lithologies conformably overlie Subunit 1 to the southeast of Ser'i Kabritu, while directly overlying the basement to the northwest of Ser'i Kabritu and at Seru Mansinga. The transition from Subunit 1 to Subunit 2 is accompanied by a significant increase in the abundance of siliciclastics, which is best illustrated by the stratigraphy of the Seru Kabritu to Blauwbaai region (Fig. 2.20). And finally, Subunit 3 Lithofacies 8 exhibits a patchy discontinuous distribution within Lithofacies 7 outcrops. Lithofacies 8 occurs only at Seru Belefa, Seru Kabritu, Seru Hermanus, Plantersrust and St. Annabaai at maximum elevations of from 10 to 30 m above modern sea level.

Depositional Environments of the Seroe Domi Formation

The limestones and sandstones of the Seroe Domi Formation were deposited in a wide-ranging sequence of deep to shallow-water marine environments. All of these environments have distinctive sedimentological characteristics

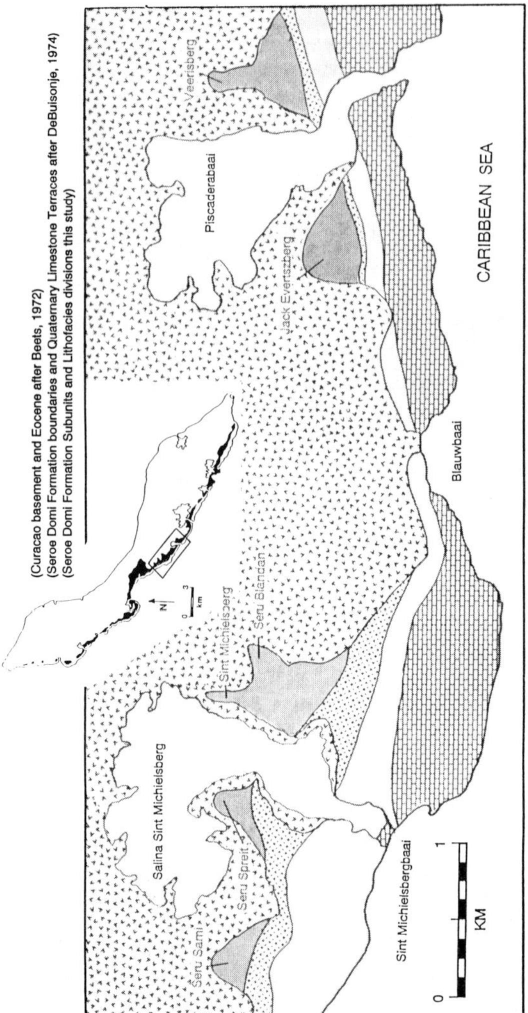

FIGURE 2.20. Geologic map of Curaçao from Seru Kabritu to Blauwbaai. The map illustrates the distribution of Subunits 1, 2, and 3 within the Seroe Domi Formation (present study), Cretaceous and Tertiary lithologies comprising the basement (after Beets, 1972), and the Quaternary Limestone Terraces (after De Buisonjé, 1974).

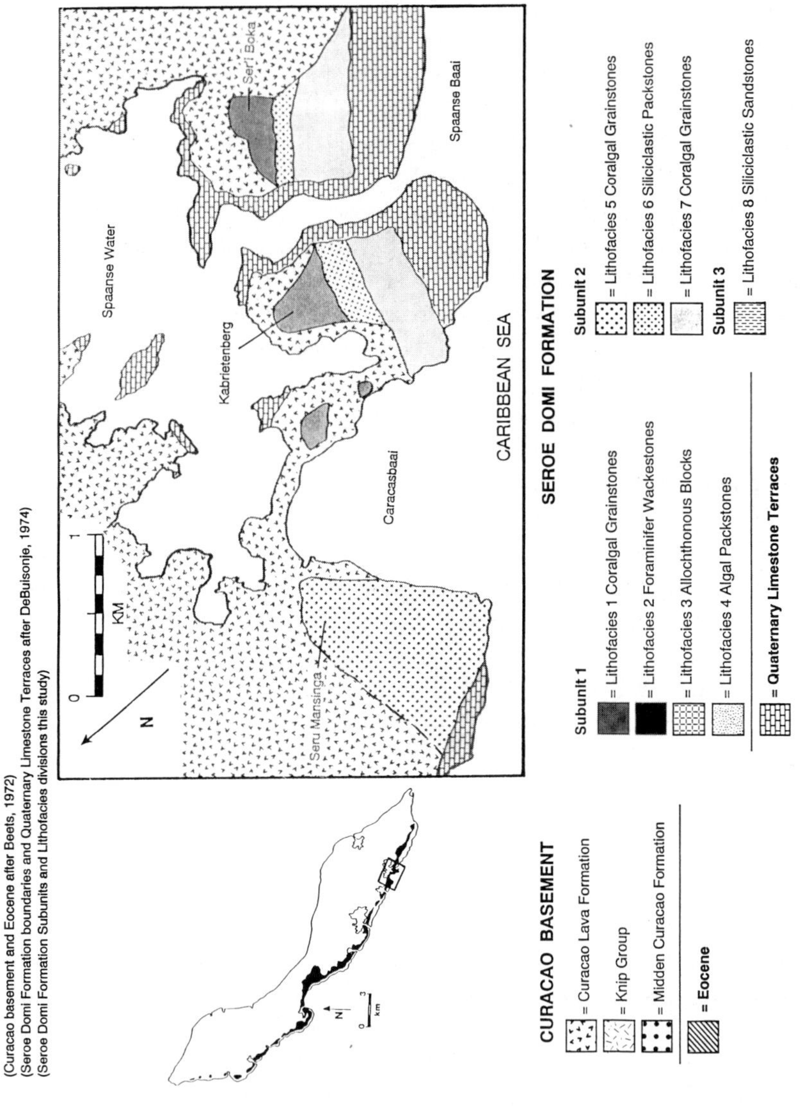

FIGURE 2.21. Geologic map of Curaçao from Ser'i Kura to Ser'i Boka. The map illustrates the distribution of Subunits 1, 2, and 3 within the Seroe Domi Formation (present study), Cretaceous and Tertiary lithologies comprising the basement (after Beets, 1972), and the Quaternary Limestone Terraces (after De Buisonjé, 1974). Note the deposition of Quaternary Limestone Terraces within channels between Seroe Domi Formation flatiron ridges.

and paleontological assemblages that permit their depositional interpretation by analogy with homologous sediments and fauna in modern marine environments on Curaçao (Bak, 1977; Bak and Luckhurst, 1980; Van Duyl, 1985) and other reef-rimmed islands and platforms throughout the Caribbean Sea and the Pacific Ocean (e.g. Adey and Burke, 1977; James and Ginsburg, 1979; MacIntyre, 1988; Hubbard *et al.*, 1990).

Terminology used in this study to describe the depositional environments of the Seroe Domi Formation follows that presented by James (1979), James and Ginsburg (1979), and Cook (1983), which combine zonations for modern and ancient reefs with terms for carbonate shelves, platform margins and basins. The distal portions of reef-rimmed shelves, where the seafloor slopes steeply toward deeper water and sedimentation results from *in situ* reef growth as well as the transportation of reef-derived debris, will be called the *reef-front* environment. This has been divided into a shallow zone (<10 m WD) and a deep zone (10 to 100 m WD) based on the water depths at which active *Acropora palmata* growth occurs on the modern shelf edge in St. Croix (Hubbard *et al.*, 1985). The term *fore-reef* will be used to describe sedimentary environments seaward of the reef-front and below the growth zone of hermatypic corals and green algae (water depths greater than approximately 100 m), which are characterized by transported sediments derived from the reef (James and Ginsburg, 1979). The fore-reef has been further subdivided into the proximal upper slope (200 to 900 m) and the distal lower slope (900 to 1300+ m; Mullins, 1985). The morphology of the seafloor at the transition from the reef front to forereef environments can vary from vertical escarpments (sedimentary by-pass margins) to gentle slopes (depositional margins; McIlreath and James, 1984). *Carbonate slope* will be used to designate environments deeper than the fore-reef , where fine-grained pelagic muds and oozes alternate with reef-derived debris (Mullins, 1979). Upper water depth limits for the fore-reef to carbonate slope transition zone in the Caribbean range from approximately 600 m in the U.S. Virgin Islands (Hubbard *et al.*, 1981) to over 2000 m in the Bahamas (Cook, 1983).

Recent carbonate deposits on Curaçao (Van Duyl, 1985) and the majority of other islands of the Greater and Lesser Antilles (Adey and Burke, 1977), are categorized as *fringing reefs*. This is because of their relatively small scale, with the break in slope and its associated prolific reef growth occurring within 0.5 to 1 km of the shoreline. The 10° to 30° inclined planar bedding of Seroe Domi Formation is interpreted as depositional slope based on the alignment of geopetals (De Buisonjé, 1974) and similarity with inclinations observed on slopes in modern and ancient reef front, fore-reef and carbonate slope environments of Curaçao and other localities (De Buisonjé, 1974; James and Ginsburg, 1979; Van Duyl, 1985; MacIntyre, 1988; Kenter, 1990). The only exceptions to this within the Seroe Domi Formation are the concave upward and landward dipping bedding of Lithofacies 5 at Seru Magasina, and the horizontal bedding of Subunit 3 Lithofacies 8.

Subunit 1 Depositional Environments

Lithofacies 1, 2, and 3: The composition, faunal assemblages and stratigraphic associations of the sediments comprising Subunit 1 Lithofacies 1, 2, and 3 suggest that they were deposited as a sequence of submarine gravity flows interbedded with pelagic carbonates in deep-water fore-reef to carbonate slope environments. The presence of the benthic foraminifers *Melonis pompiloides* (Fitchell and Moll) and *Laticarinina pauperata* (Parker and Jones) in Lithofacies 2 suggests deposition took place at water depths of at least 500 m (Melillo, pers. comm.) and perhaps up to 2,000 m (Pflum *et al.*, 1976 p. 28). Deep-water deposition is also suggested by the similarity in the high abundance of pelagic foraminifers and muds in Lithofacies 2 wackestones compared to foraminifer-rich pelagic muds recovered from modern deep-water platform-margin environments adjacent to Belize (James and Ginsburg, 1979), the Bahamas (Cook, 1983), Barbados (MacIntyre, 1967), and St. Croix (Hubbard *et al.*, 1981). Deep water deposition may also be implied by the trace fossils observed in Lithofacies 2 at Fort Nassau. Although not systematically identified in this study, these trace fossils are similar in two-dimensional form to specimens described from the deep-water *Zoophycos* and *Nerites* ichnofacies (Seilacher, 1978; Ekdale *et al.*, 1984).

The grainstones, packstones and allochthonous blocks comprising Lithofacies 1 and 3 are also interpreted to have been deposited in deep-water fore-reef to carbonate slope environments based on: 1) their interbedding with Lithofacies 2; 2) similarity to modern fore-reef and carbonate slope sediments; and 3) high abundance of pelagic foraminifers with respect to benthic foraminifers. Similar foraminifer-rich pelagic muds interbedded with shallow-water platform-derived carbonate gravity flows have been observed in modern platform apron deposits at water depths as great as 4000 m off of St. Croix (Hubbard *et al.*, 1981) and from 600 to over 2,500 m in the Bahamas (Mullins, 1979; Cook, 1983; Mullins, 1983). Based on analogy with these modern deposits, the Lithofacies 1 to Lithofacies 2 facies change may represent either: 1) increases in relative sea level serving to spatially removed the depositional area from the platform sediment source; 2) changes in the relative frictional and cohesive strength of the gravity flows, thus changing the extent of lateral transport along the slope; 3) progradation of the reef front; or 4) a combination of these effects triggered tectonically or eustatically (Cook, 1983; Austin *et al.*, 1988; Kenter, 1990). The 3:1 ratio of pelagic to benthic foraminifers observed in Lithofacies 1 is consistent with ratios in modern deep-water sediments at water depths from 500 m to over 2,000 m (Pflum *et al.*, 1976). Lithofacies 3 allochthonous blocks are interpreted to represent sporadic rockfalls from steep up-dip slopes, in which the impact of the falling block distorted the sedimentary beds on which it landed. Similar allochthonous blocks are common in both modern and ancient reef-front, fore-reef and carbonate slope environments (James and Ginsburg, 1979; McIlreath and James, 1982; Hubbard *et al.*, 1985).

The paleobathymetric depositional evidence for Lithofacies 1, 2, and 3, in conjunction with the shallow-water (< 100 m) normal marine origin and conglomeratic nature of the Lithofacies 1 and 3 sediments, suggests that they were transported from shallow to deep water depositional environments as multiple small scale sediment gravity flows. Transport mechanisms active during deposition of Lithofacies 1, 2, and 3 were presumably similar to those observed in deep-water carbonate deposits in lower fore-reef slope environments in the Bahamas (Crevello and Schlager, 1980; Cook, 1983; Mullins, 1985; Austin *et al.*, 1988). The non-graded, coarse grained texture, small percentage of mud, and 1 m-thick bedding of Lithofacies 1 are consistent with deposition from the combined effects of grain flows and turbidity flows. The absence of slump structures and the 30° inclination observed in Lithofacies 1 deposits are also consistent with steep slope sediments in modern Bahamian fore-reef environments (Crevello and Schlager, 1980; Hine, 1983; Mullins, 1983, 1985).

Lithofacies 4: The red algal growth forms, sedimentologic composition, seaward dip, and conformably overlying Lithofacies 7 grainstones at Kabrietenberg suggest that Lithofacies 4 deposition occurred in fore-reef seafloor environments similar to those of Lithofacies 1. In addition, the coarse grain size, poor sorting, and algal-dominated composition of Lithofacies 4 sediments are similar to transported sediments dredged from the fore-reef in

Hawaii and Florida (Bosence, 1991; Adey *et al.*, 1982). Recent rhodoliths occur in a variety of back reef, reef flat and reef-front environments with growth forms that range from free-living nodules to *in situ* reef frameworks (Bossellini and Ginsburg, 1971; Adey and McIntyre, 1973; Bosence, 1983, 1984). Living tropical rhodoliths are also restricted to water depths of less than approximately 100 m (Adey *et al.*, 1982), and exhibit morphological plasticity in response to physical disturbance from wave energy and/or bioturbation (Bosellini and Ginsburg, 1971; Bosence and Pedley, 1982). The densely branched to laminar spheroidal form of the thalli (coralline skeletons) in Lithofacies 4 rhodoliths suggests growth in relatively high wave energy environments in which the thalli were repeatedly overturned to form concentric laminar growth layers (Bosellini and Ginsburg, 1971; Bosence, 1983, 1985). However, this form may also represent rolling during multiple episodes of down-slope transportation. The relatively low abundance of corals and spheroidal rhodolith morphologies of Lithofacies 4 sediments (Fig. 2.3) suggest that they may have been derived from environments similar to those associated with algal reefs in St. Croix (Bosence, 1983).

Subunit 2 Depositional Environments

Lithofacies 5: Lithofacies 5 bedding characteristics and faunal assemblages are suggestive of deposition in shelf to reef front environments. The presence of *in situ* corals and aligned internal sediment geopetals suggest that Lithofacies 5 sediments have not been transported. In addition, the corals and associated sediments of Lithofacies 5 are similar to those in shallow-water (<30 m) moderate wave energy reef front and backreef environments on the modern shelves of Curaçao (Bak, 1977).

De Buisonjé and Zonneveld (1976) have interpreted a bathymetric high at the end of an elongate seafloor scarp just seaward of Caracasbaai (the bay between Locations 38 and 39 in Fig. 1.8) as a large block of the Seroe Domi Formation that dislodged and slid downslope during the Pleistocene. The distorted landward-dipping bedding and extensive vertical fracturing of Seru Mansinga observed in this study are consistent with tectonic activity and/or gravity slumping in the area. The direction and magnitude of the proposed Caracasbaai slide also suggests that the Lithofacies 5 deposits could have been transported downslope as a coherent slide block, thus preserving the coral growth positions and geopetal orientations.

Lithofacies 6: The abundance of siliciclastic sands and steeply inclined bedding suggest that Lithofacies 6 sediments were derived from the Curaçao basement and deposited as submarine gravity flows in reef-front to fore-reef environments. The high 3:1 ratio of benthic to pelagic foraminifers suggests deposition in water depths of less than 200 m (Pflum *et al.*, 1976), which is significantly more shallow than Lithofacies 1 depositional environments. The distinctly low abundance of corals in Lithofacies 6 may represent either the prevention of coral growth due to the high influx of siliciclastics (Hubbard, 1986), or depositional water depths below the coral photic zone (<100 m WD, James and Ginsburg, 1979). However, coral debris derived from higher parts of the slope would be expected in these deposits if changes in water depth were the sole cause. Therefore, the siliciclastic-rich nature of Lithofacies 6 packstones suggest that the exclusion of corals was probably due to intense siliciclastic sedimentation, which is common on modern shelves throughout the Caribbean (Moorelock *et al.*, 1983; Hubbard, 1986;). The siliciclastic grains in Lithofacies 6 are similar in composition to the Curaçao basement, which suggests that they were derived from erosion of the Curaçao Lava Formation, Midden Curaçao Formation and the Knip Group.

Lithofacies 7: Lithofacies 7 is composed of sediments transported from shallow shelf source areas mixed with *in situ* skeletal growth on the reef front that trapped and bound the transported sediments. The coral assemblages and associated sediments of Lithofacies 7 (Table 2.3) are comparable with fauna observed in moderate wave energy reef front environments on the modern shelves of Curaçao (Bak, 1977). In addition, the general orientation of elongate coral fronds such as *A. panamensis* parallel to the direction of dip are suggestive of off-shelf water currents and/or sediment movement (De Buisonjé, 1974). The extent to which carbonate sediments have been transported in reef-front environments is generally difficult to determine because: 1) in-place coral and algae framework builders, which generally cover less than 30% of the seafloor in modern reef front deposits, are often physically overturned so that little *in situ* fabric is preserved; 2) the distance over which transportation takes place is small, causing relatively little abrasion; and 3) sedimentation in this environment is a continually changing admixture of these autochthonous and allochthonous processes (Hubbard *et al.*, 1985, 1990). *Agaracia lamarcki* boundstones within Lithofacies 7 represent local areas of prolific *Agaracia* growth analogous to those occurring on the modern reef at water depths of 15 to 50 m (Bak, 1977; Van Duyl, 1986). These *Agaracia* boundstones also often thicken seaward, suggesting progradation of the reef-front. The diminutive brachiopod *T. barretti* observed at St. Annabaai is one of a functionally distinct group of brachiopods that permanently cement their brachial valve to the undersurfaces of *in situ* corals in reef-front environments at less than 100 m water depth on modern Caribbean fringing reefs (Jackson *et al.*, 1971; Fouke and LaBarbera, 1986). The clustered distribution and high densities of the brachiopods and the associated bryozoan, serpulid and foraminifer epibenthos found with *T. barretti* on the undersurfaces of *Agaracia lamarcki* in the Lithofacies 7 are identical to those observed in modern reefs. This is further suggestive of deposition in shallow-water reef-front environments. It is important to recognize that the modern reef front on Curaçao has a precipitous slope of nearly 50° to 90° in many places, while the maximum slopes in the Seroe Domi Formation are only about 30° (Bak, 1977; Van Duyl, 1986). This difference in shelf profile between the modern and ancient may be a function of either the distance of contemporaneous carbonate production from the deep-water shelf break or the extent of coral development. A narrow shelf would not have room to accommodate progradation due to off-shelf sediment transport, and would therefore preserve only vertical aggradational deposits. In addition, extensive reef development would steepen the reef-front profile (James and Ginsburg, 1979).

Subunit 3 Depositional Environments

Lithofacies 8: As is the case for Lithofacies 6 packstones, the siliciclastics comprising Lithofacies 8 are identical in composition to lithologies comprising the Curaçao Lava Formation, Midden Curaçao Formation and the Knip Group. This suggests that the siliciclastics were sourced from erosion of the basement of the island. The truncated bedding in Lithofacies 8 suggests that strong currents moved within the Lithofacies 7 chutes and caves, resulting in multiple filling and erosional events. The abundance of abraded and rounded clasts of the coral *A. palamata* in Lithofacies 8 suggests that this extensive lateral transport took place in the marine environment and was therefore generated by either currents, waves or storms. The rounding of the siliciclastic grains also suggests extensive marine transportation.

The deposition of Lithofacies 8 requires that the source area of the siliciclastics was equal to or greater than the present-day elevation of the deposit, which is at 40 m. Therefore, the basement lithologies have been eroded at least 40 m since Lithofacies 8 deposition. The process that formed the stratiform chutes within Lithofacies 7 remains uncertain. It is possible that they resulted from a combination of submarine erosion, subaerial erosion and meteoric karstic processes. However, regardless of their formation, the stratigraphic sequence requires that these chutes and caves were empty caverns prior to being filled with siliciclastic sediments.

Conclusions

The Seroe Domi Formation is composed of three lithologically distinct and mapable divisions called Subunits 1, 2, and 3. Subunit 1, the basal part of the Seroe Domi Formation, is a 30 to 100 m-thick sequence of Lithofacies 1 coralgal packstones and grainstones interbedded with Lithofacies 2 foraminifer wackestones and Lithofacies 3 allochthonous grainstone blocks. Subunit 1 also contains a pervasively dolomitized Lithofacies 4 coralline algal grainstone. Subunit 2 comprises the upper 250 to 300 m of the Seroe Domi, and is composed Lithofacies 5 grainstones, as well as siliciclastic-rich Lithofacies 6 packstones overlain by Lithofacies 7 coralgal grainstones and boundstones. Subunit 3 is composed of horizontally bedded Lithofacies 8 siliciclastic sandstones and conglomerates, deposited in preferentially eroded stratiform chutes and caves within Subunit 2 Lithofacies 7.

Subunit 1 is aerially restricted to the southeastern half of Curaçao, with Lithofacies 1, 2 and 3 cropping out northwest of Seru Mansinga and Lithofacies 4 distributed to the southeast. Subunit 2 Lithofacies 5 is stratigraphically isolated from other Seroe Domi Formation deposits, and is found only at Seru Mansinga. Conversely, Subunit 2 Lithofacies 6 and 7 crop out over the northwestern three-fourths of the island, onlapping Subunit 1 in the southeast and directly overlying the basement to the northwest. Subunit 3 Lithofacies 8 has a discontinuous distribution within Lithofacies 7 outcrops.

Subunit 1 Lithofacies 1, 2, and 3 are a sequence of carbonate gravity flows interbedded with pelagic oozes deposited in deep-water fore-reef to carbonate slope environments. The presence of the benthic foraminifer *Melonis pompiloides* (Fitchell and Moll) in Lithofacies 2 wackestones, and the similarity of Subunit 1 deposits with modern deep-water carbonates, suggests depositional water depths of at least 500 m. The depositional setting of Lithofacies 4 is uncertain due to fabric-destructive dolomitization, and may represent deposition at a variety of marine fore-reef paleobathymetries. The siliciclastics in Subunit 2 Lithofacies 6 are interpreted to have been sourced from the erosion of the basement during subaerial exposure and deposited in shallow fore-reef to reef-front marine environments. The presence of *in situ* reefs in Subunit 2 Lithofacies 5 and 7 indicates depositional water depths of less than 100 m within the photic zone. Subunit 3 Lithofacies 8 sandstones and conglomerates are interpreted to have been deposited in shallow marine to fluvial environments.

CHAPTER 3

CHRONOSTRATIGRAPHY AND DEPOSITIONAL HISTORY

In this chapter biostratigraphic and $^{87}Sr/^{86}Sr$ age dating techniques are applied to the lithologic sequence presented in Chapter 2, in order to establish the depositional history of the Seroe Domi Formation.

Chronostratigraphy of the Seroe Domi Formation

Biostratigraphy

Reevaluation of Seroe Domi Formation scleractinian coral and pelagic foraminifer assemblages in the present study has significantly expanded and refined the original biostratigraphy presented by De Buisonjé (1974). The most diagnostic biostratigraphic indicator occurs in Subunit 1 Lithofacies 2 wackestones sampled from 12 m above the basement contact at Seru Blandan (Fig. 3.1). At these horizons, the presence of the pelagic foraminifer *Praeorbulina glomerosa circularis* (Blow; Fig. 2.7) coupled with the absence of the genus *Orbulina*, suggests an early Middle Miocene depositional age of 16.2 to 15.2 Ma (after Berggren *et al.*, 1985; Melillo, 1985; Fig. 3.2). The presence of *Globorotalia foshi peripheroronda* and *Globorotalia mayeri* (Figs. 2.7 and 2.8) in these wackestones further supports a depositional age of early Middle Miocene (Berggren *et al.*, 1985; Melillo, 1985). In addition, Lithofacies 2 contains no evidence of resedimentation and is interbedded with Lithofacies 1 and 3 grainstones. Therefore Lithofacies 1 and 3 are also Middle Miocene in age. The biostratigraphic range zones of the corals observed in Lithofacies 1, 2, and 3 extend throughout the Miocene and Pliocene (Frost, 1977; Budd *et al.*, in press; Fig. 3.3). Therefore, although consistent with the foraminifer-derived ages, the coral biostratigraphy does not provide further chronologic resolution.

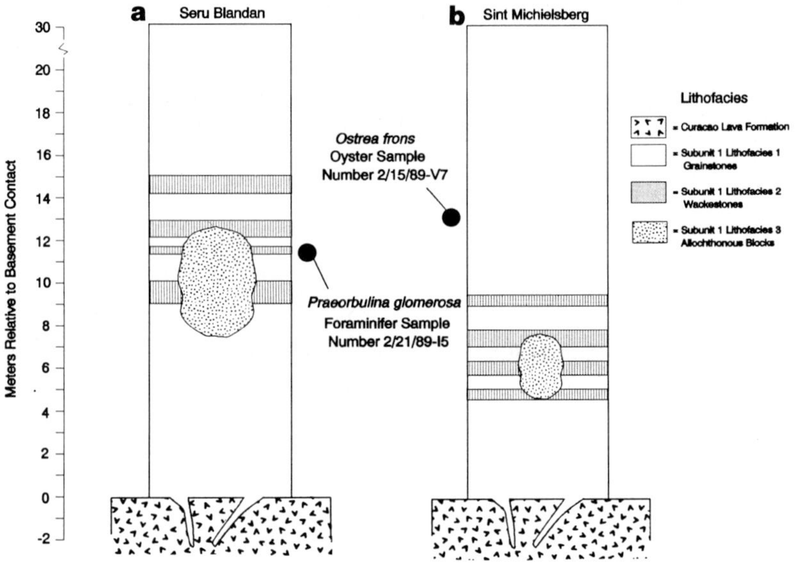

FIGURE 3.1. Geographic and stratigraphic position of foraminifer and oyster skeletal material used for chronostratigraphic analyses of the Seroe Domi Formation. Stratigraphic sections for Seru Blandan (a) and St. Michielsberg (b) illustrating the relative positions of the *Praeorbulina glomerosa* foraminifer specimens from Seru Blandan (sample number 2/21/89-I5) and the *Ostrea frons* oyster specimens from St. Michielsberg (number 2/15/89-V7).

The fabric-destructive dolomitization of Lithofacies 4 has prevented the identification of foraminifers in this deposit. As with Lithofacies 1, 2, and 3, Lithofacies 4 coral range zones do not delimit the age of deposition more precisely than Miocene to Pliocene (Fig. 3.3). However, Lithofacies 4 deposits are stratigraphically overlain by Lithofacies 6 and 7 deposits at Kabrietenberg and Seru Boca. This constrains the minimum age of Lithofacies 4 to being older than Subunit 2.

Although Subunits 2 and 3 contain no well-preserved foraminifers, their coral assemblages and stratigraphic relationships have provided valuable chronologic resolution. Lithofacies 5 contains the corals *Leptoseris sp.*, *Stylophora canalis*, *Dichocoenia stokesii*, and *Montastrea limbata*, whose overlapping ranges suggest a depositional age of Late Miocene to Early Pliocene (Fig. 3.3). A similar assemblage occurs in Lithofacies 7 grainstones, with the addition of *Diploria clivosa*, *Montastrea annularis*, and *M. imperatoris*, which also suggests

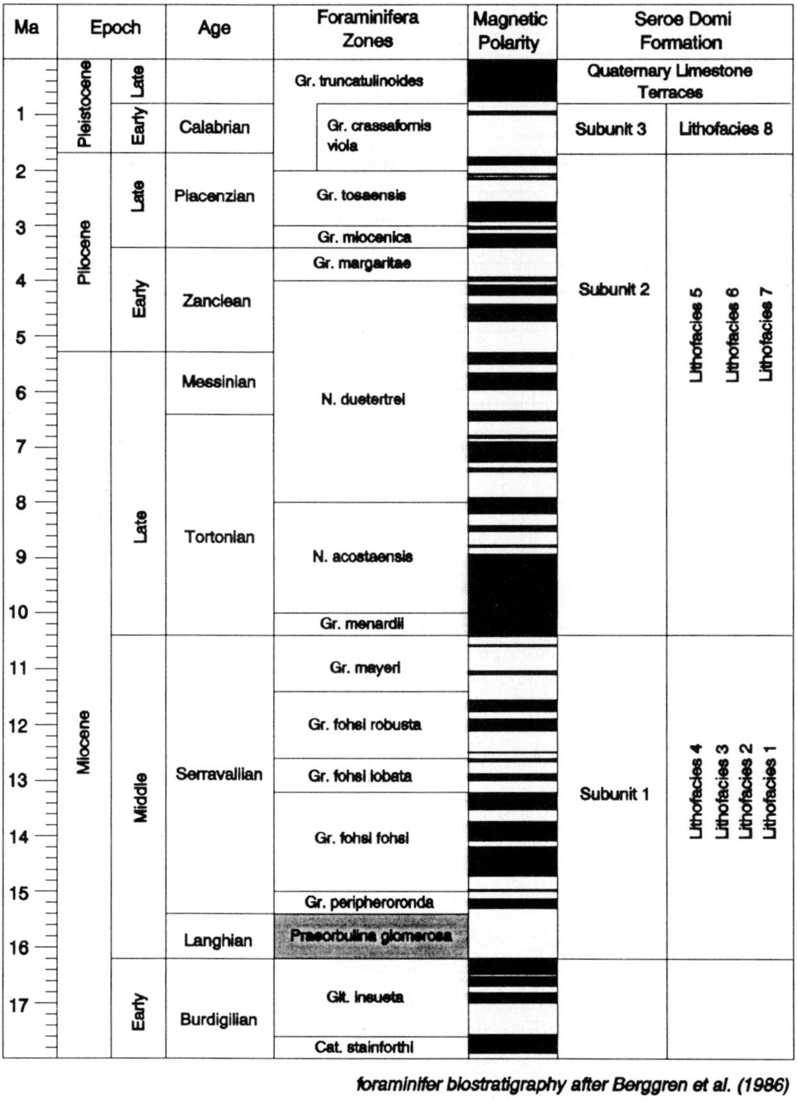

FIGURE 3.2. Mio-Plio-Pleistocene paleomagnetic reversal stratigraphy, with correlated foraminifer biostratigraphy and Seroe Domi Formation stratigraphy. Reversal and foraminifer stratigraphy from Berggren *et al.* (1986).

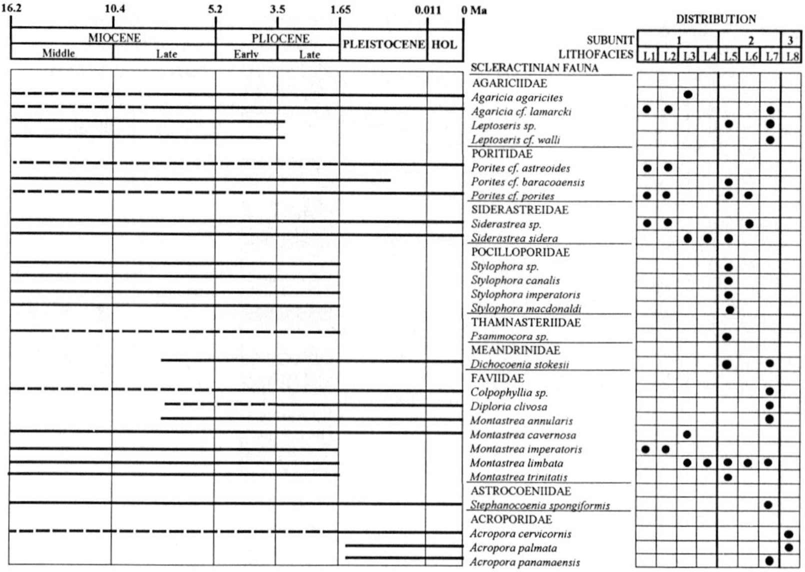

Biostratigraphic ranges after Frost (1977) and Budd *et al.* (in press).

FIGURE 3.3. Scleractinian coral faunal assemblages and biostratigraphic ranges observed in the Seroe Domi Formation on Curaçao. Caribbean scleractinian biostratigraphic range zones after Frost (1977) and Budd *et al.* (in press).

deposition during the Late Miocene to Early Pliocene (Fig. 3.3). Conversely, Lithofacies 6 packstones contain *M. imperatoris*, *Porites cf. porites*, and *Siderastrea sidera*, which provide only a coarse age resolution of Miocene to Pliocene (Fig. 3.3). Therefore the relative age of Lithofacies 6 has been determined in the St. Michielsberg region, where it is underlain by Subunit 1 and overlain by Lithofacies 7 and suggests an age of late Middle Miocene to Early Pliocene. And finally, Subunit 3 Lithofacies 8 contains fragments of *Acropora palmata* and is onlapped by Quaternary Limestone Terraces. This suggests that Subunit 3 was deposited during the Early Pleistocene (Fig. 3.3).

Sr Isotope Geochronology

Strontium isotope analyses of marine carbonates from Deep Sea Drilling Program (DSDP) and Ocean Drilling Program (ODP) core sections have documented an increase of approximately 0.0005 in the $^{87}Sr/^{86}Sr$ of seawater since the Early Miocene (Burke *et al.*, 1982; Koepnick *et al.*, 1985; DePaolo, 1986; Hess *et al.*, 1986; Hodell *et al.*, 1989, 1990, 1991; Miller *et al.*, 1991; Beets, 1991, 1992). The correlation of $^{87}Sr/^{86}Sr$ in unaltered marine skeletal material with this DSDP-ODP seawater $^{87}Sr/^{86}Sr$ curve permits high-resolution age dating of deposition (e.g. Ludwig *et al.*, 1988; Rundberg and Smalley, 1989; Saller and Koepnick, 1990; DePaolo and Finger, 1991) and diagenesis (e.g. Saller, 1984; Richter and DePaolo; 1987; Swart *et al.*, 1987; Banner *et al.*, 1988; Saller and Koepnick, 1990; Vahrenkamp and Swart, 1987, 1988) when integrated with other available geological and geochemical evidence (Halley and Ludwig, 1989). Plateaus in the seawater $^{87}Sr/^{86}Sr$ curve during the Tertiary prevent chronologic resolution *within* the Late Miocene and Pliocene. However, it is possible to distinguish *between* the Middle Miocene, Late Miocene, Pliocene and Pleistocene (Hodell *et al.*, 1991; Beets, 1992). Therefore, Sr isotope geochronology is applied here in combination with several other independent lines of geological and geochemical evidence to determine the depositional history of the Seroe Domi Formation.

The $^{87}Sr/^{86}Sr$ compositions of oyster and brachiopod shells from the Seroe Domi Formation have been compared with co-occurring calcite cements as well as modern shells and modern groundwaters. This has permitted quantitative reconstructions of the original geochemical compositions of Seroe Domi Formation shells for use in $^{87}Sr/^{86}Sr$ age dating. Explanation of the microsampling procedures, geochemical analyses and quantitative modelling techniques are presented in Fouke (1993). Shells of the oyster *Ostrea frons* and the brachiopod *Thecidellina barretti* were collected from Subunit 1 Lithofacies 1 and Subunit 2 Lithofacies 6 and 7 limestones of the Seroe Domi Formation (Figs. 3.1 and 3.4). Living specimens of the brachiopod *T. barretti* were collected on SCUBA at 15 m water depth off the southern coast of Curaçao near Piscadera Bay from cryptic hardgrounds within constructional crevices of the reef-front (offshore from Locality 29 in Fig. 1.8). Living specimens of *O. frons* were collected at 1 m water depth from the mouth of the Piscadera Bay channel just landward of the brachiopod locality. The location, sample number, taxonomic identification, degree of textural alteration, and geochemical composition of each sample is presented in Table 3.1.

Petrographic criteria were established in order to consistently categorize Seroe Domi Formation oyster and brachiopod shells according to the extent to which they had been texturally altered. Texturally *unaltered* or *pristine* speci-

Sampling Locality	Sample Number	Textural Alteration	Mg ppm	Mg mol%	Sr ppm	1/Sr ppm	Sr/Ca 1000x	Mn ppm	Fe ppm	$\delta^{18}O$ PDB	$\delta^{13}C$ PDB	$^{87}Sr/^{86}Sr$
		Standard Deviation	85	0.04	6	0.17		7	51	0.10	0.20	0.000020
		Detection Limit	69	0.03	2	0.50		22	54			
Subunit 1 Lithofacies 1 Oyster *Ostrea frons*												
St Michielsberg	2/15/89-V7	None	2001	0.8	233	0.004	0.59	43	57	-1.90	-3.33	0.708781
St Michielsberg	2/15/89-V2	Partial	1697	0.7	117	0.009	0.29	73	70	-2.06	-4.37	0.708807
St Michielsberg	2/15/89-V7	Partial	1967	0.8	200	0.005	0.50	82	93	-1.67	-3.22	0.708799
St Michielsberg	2/15/89-V7	Complete	1351	0.6	53	0.019	0.13	137	78	-1.95	-4.31	0.708832
St Michielsberg	2/15/89-V7	Complete	1694	0.7	92	0.011	0.23	142	125	-1.78	-4.26	
St Michielsberg	2/15/89-V7	Complete	1408	0.6	110	0.009	0.28	149	161	-1.58	-3.57	
St Michielsberg	2/15/89-V7	None	1064	0.4	112	0.009	0.28	86	57	-1.78	-3.77	
St Michielsberg	2/15/89-V7	None	1612	0.7	144	0.007	0.36	63	64			
St Michielsberg	2/15/89-V7	Partial	1231	0.5	95	0.011	0.24	115	57			
St Michielsberg	2/15/89-V7	Partial	1328	0.5	122	0.008	0.31	64	67			
St Michielsberg	2/15/89-V7	None	1320	0.5	37	0.027	0.09	58	*bd*			
St Michielsberg	2/15/89-V7	Complete	1573	0.6	33	0.030	0.08	205	*bd*			
Average Composition			1521	0.6	112	0.012	0.28	101	83	-1.81	-3.95	0.708805
Subunit 2 Lithofacies 6 Oyster *Ostrea frons*												
Seru Spreit	9/16/90-IIB	None	2052	0.8	779	0.001	1.96	*bd*	*bd*	-1.34	-2.75	0.709008
Seru Spreit	9/16/90-IIB	None	1659	0.7	551	0.002	1.39	*bd*	*bd*			
Seru Spreit	9/16/90-IIB	Partial	1728	0.7	339	0.003	0.85	43	62			
Seru Spreit	9/16/90-IIB	Complete	1763	0.7	397	0.003	1.00	231	*bd*			
Seru Spreit	9/16/90-IIB	Partial	1771	0.7	594	0.002	1.49	32	*bd*			
Average Composition			1795	0.7	532	0.002	1.34	102	*bd*	-1.34	-2.75	0.709008
Subunit 2 Lithofacies 7 Oyster *Ostrea frons*												
St Annabaai	9/15/90-1A	None	2697	1.1	700	0.001	1.77	*bd*	*bd*	-1.27	-2.72	0.709045
St Annabaai	9/15/90-1A	Partial	2685	1.1	576	0.002	1.45	58	*bd*	-1.53	-2.86	0.709059
St Annabaai	9/15/90-1A	Complete	1269	0.5	263	0.004	0.66	119	*bd*	-1.62	-3.45	0.709063
St Annabaai	9/15/90-1A	Complete	2424	1.0	529	0.002	1.33	121	*bd*	-1.52	-2.97	
St Annabaai	9/15/90-1A	Complete	2322	1.0	588	0.002	1.48	93	*bd*			
St Annabaai	9/15/90-1A	None	1774	0.7	740	0.001	1.86	25	*bd*			
St Annabaai	9/15/90-1A	Partial	1836	0.8	446	0.002	1.12	30	*bd*			
Average Composition			2144	0.9	549	0.002	1.38	74	*bd*	-1.49	-3.00	0.709056
Subunit 2 Lithofacies 7 Brachiopod *Thecidellina barretti*												
St. Annabaai	9/15/90-1B	None	8502	3.5	975	0.001	2.52	*bd*	*bd*	-0.88	-0.88	0.709035
St. Annabaai	9/15/90-1B	Partial	7314	3.0	214	0.005	0.55	69	*bd*	-1.00	-1.45	0.709044
St. Annabaai	9/15/90-1B	Complete	4624	1.9	94	0.011	0.24	207	95	-0.72	-3.83	0.709050
St. Annabaai	9/15/90-1B	Complete	4776	2.0	89	0.011	0.23	248	250	-0.78	-2.99	
St. Annabaai	9/15/90-1B	Partial	4723	1.9	238	0.004	0.61	45	*bd*	-0.60	-2.13	
St. Annabaai	9/15/90-1B	Partial	4794	2.0	491	0.002	1.25	55	*bd*	-0.38	-1.46	
St. Annabaai	9/15/90-1B	Partial	5913	2.4	328	0.003	0.84	27	*bd*			
St. Annabaai	9/15/90-1B	Partial	6841	2.8	711	0.001	1.83	72	*bd*			
St. Annabaai	9/15/90-1B	Complete	8671	3.6	404	0.002	1.05	255	382			
St. Annabaai	9/15/90-1B	Complete	11068	4.6	582	0.002	1.52	172	185			
St. Annabaai	9/15/90-1B	Complete	12647	5.2	459	0.002	1.21	344	314			
St. Annabaai	9/15/90-1B	Partial	9816	4.0	573	0.002	1.49	36	135			
St. Annabaai	9/15/90-1B	None	6583	2.7	664	0.002	1.70	23	*bd*			
St. Annabaai	9/15/90-1B	Partial	8854	3.6	236	0.004	0.61	125	*bd*			
St. Annabaai	9/15/90-1B	Complete	8859	3.6	326	0.003	0.85	154	*bd*			
St. Annabaai	9/15/90-1B	Partial	5800	2.4	1141	0.001	2.92	*bd*	*bd*			
St. Annabaai	9/15/90-1B	Partial	4941	2.0	1082	0.001	2.76	*bd*	*bd*			
St. Annabaai	9/15/90-1B	Partial	8638	3.6	335	0.003	0.87	94	181			
St. Annabaai	9/15/90-1B	Complete	12626	5.2	708	0.001	1.87	198	69			
St. Annabaai	9/15/90-1B	Complete	9907	4.1	240	0.004	0.62	163	326			
Average Composition			7782	3	483	0.003	1.25	139	215	-0.62	-2.60	0.709043
Recent Curacao Oyster *Ostrea frons*												
Piscaderabaai	6/25/88-II1	None	5653	2.3	567	0.002	1.45	*bd*	*bd*	-0.12	-0.15	
Piscaderabaai	6/25/88-II1	None	5484	2.3	430	0.002	1.10	*bd*	*bd*	-0.54	-0.33	
Piscaderabaai	6/25/88-II1	None	5533	2.3	808	0.001	2.07	*bd*	*bd*			
Piscaderabaai	6/25/88-II1	None	5439	2.2	748	0.001	1.91	*bd*	*bd*			
Average Composition			5527	2	638	0.002	1.63	*bd*	*bd*	-0.33	-0.24	
Recent Curacao Brachiopod *Thecidellina barretti*												
Piscaderabaai	6/25/88-I1	None	7094	2.9	1411	0.001	2.52	*bd*	*bd*	-0.83	1.65	
Piscaderabaai	6/25/88-I1	None	10382	4.3	1411	0.001	3.68	*bd*	*bd*	-0.67	1.31	
Piscaderabaai	6/25/88-I1	None	4848	2.0	1216	0.001	3.10	*bd*	*bd*	-0.91	1.94	
Piscaderabaai	6/25/88-I1	None	7388	3.0	1257	0.001	3.24	*bd*	*bd*			
Piscaderabaai	6/25/88-I1	None	7635	3.1	1472	0.001	3.80	*bd*	*bd*			
Piscaderabaai	6/25/88-I1	None	9944	4.1	1481	0.001	3.86	*bd*	*bd*			
Piscaderabaai	6/25/88-I1	None	17333	7.1	1574	0.001	4.23	*bd*	*bd*			
Average Composition			9232	4	1403	0.001	3.49	*bd*	*bd*	-0.80	1.63	

bd = below detection

Ca, Mg, and Fe determined by electron probe microanalysis (EPMA).

Sr and Mn determined by direct coupled plasma (DCP) analysis.

TABLE 3.1. The geochemical compositions of Recent and Seroe Domi Formation *Ostrea frons* oyster and *Thecidellina barretti* brachiopod shells.

FIGURE 3.4. Geographic and stratigraphic position of Subunit 2 Lithofacies 7 oyster and brachiopod skeletal material used for chronostratigraphic analyses of the Seroe Domi Formation. A. Western bank of the channel at St. Annabaai, illustrating eroded troughs in Subunit 2 Lithofacies 7 coralgal grainstones (L7) filled with Subunit 3 Lithofacies 8 siliciclastic sandstones and conglomerates (L8). B. Enlargement of area indicated by the arrow on photograph A. *Ostrea frons* oyster sample number 9/15/90-IA was collected from the lithologies pointed at by the pick of the rock hammer. C. Enlargement of area indicated by the arrow on photograph B. Arrow points to an *in situ Thecidellina barretti* that is part of sample number 9/15/90-IB.

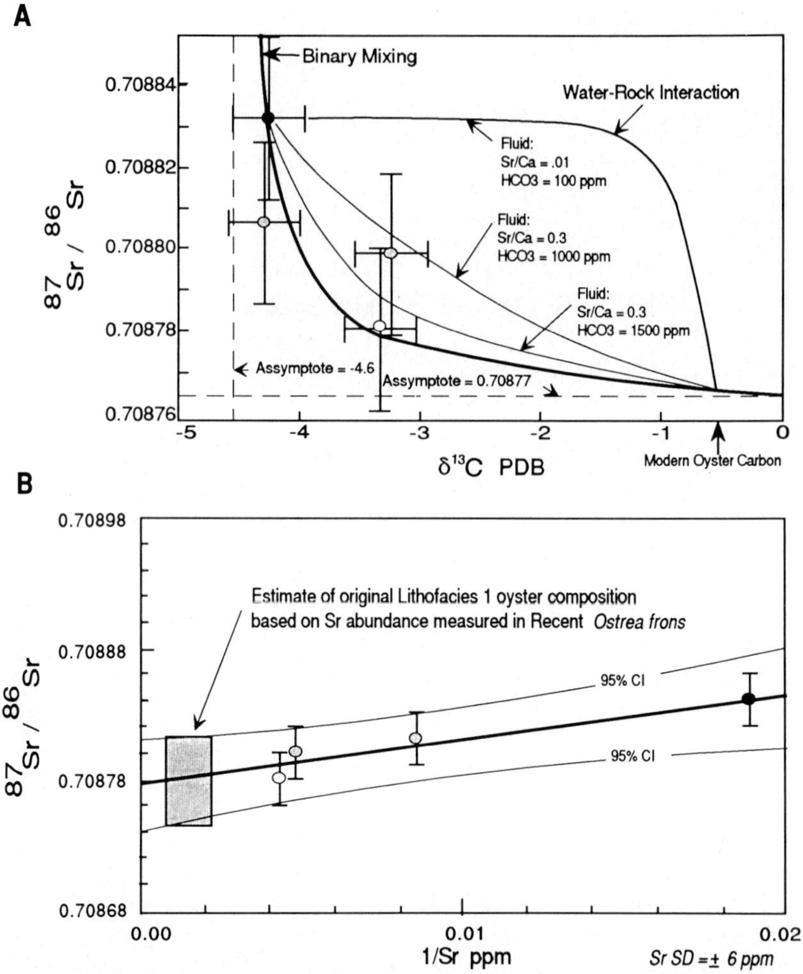

FIGURE 3.5. Compositional covariations of $^{87}Sr/^{86}Sr$, $\delta^{13}C$ and Sr in texturally unaltered, partially altered, and completely altered Subunit 1 Lithofacies 1 *Ostrea frons* shells. Recent and Seroe Domi Formation *O. frons* compositions presented in Table 3.1. Seawater and freshwater fluid compositions used in these calculations are presented in Table 3.4. Binary mixing trajectories and water-rock interaction reaction pathways were modelled using equations presented in Banner and Hanson (1991) and Langmuir *et al.* (1978). A. Cross-plot of $\delta^{13}C$ versus $^{87}Sr/^{86}Sr$. B. Companion cross-plot to (C) of 1/Sr versus $^{87}Sr/^{86}Sr$.

mens are shells that are non-cathodoluminescent (CL), and exhibit well-preserved fabrics identical to those observed in modern shells. *Partially altered*

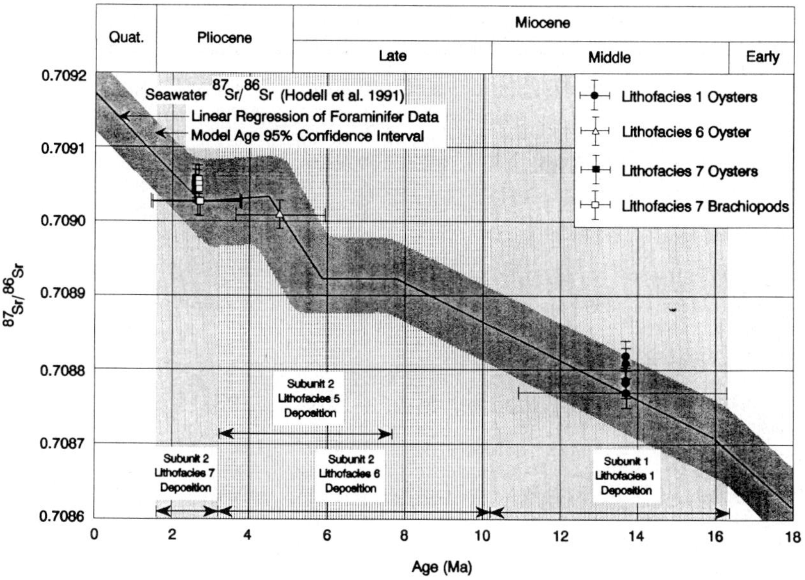

FIGURE 3.6. Plot of Seroe Domi Formation skeletal $^{87}Sr/^{86}Sr$ versus Hodell *et al.* (1991) model ages.

specimens include shells with localized areas of mottled CL, and patchy distributions of smoothed to irregular pits and holes caused by dissolution within the original shell fabric. *Completely altered* shells exhibit pervasive bright to mottled CL and extensive dissolution fabrics, with only traces of the original shell fabric. Unaltered shells from both the base (Subunit 1 Lithofacies 1) and top (Subunit 2 Lithofacies 7) of the Seroe Domi Formation are depleted in $\delta^{18}O$, $\delta^{13}C$ and Sr and enriched in Mn, Mg, and Fe with respect to the equivalent modern shells. This suggests that texturally unaltered as well as texturally altered shells experienced meteoric diagenesis that partially reset their $^{87}Sr/^{86}Sr$ signatures.

Subunit 1 Lithofacies 1 *O. frons* $^{87}Sr/^{86}Sr$ versus $\delta^{13}C$ and 1/Sr covariations closely correlate with modelled trajectories representing mechanical mixtures of skeletal calcite and replacement calcite, rather than trajectories representing water-rock interaction (Fig. 3.5). The hyperbolic asymptotes of mixing curves fitted with the $^{87}Sr/^{86}Sr$ versus $\delta^{13}C$ shell compositions suggest that the shells had an initial $^{87}Sr/^{86}Sr$ of 0.70879 ($\pm$0.00002; Fig. 3.5). This implies a Middle Miocene depositional age of 13.8 $\pm$1.4 Ma (Table 3.2;

			$^{87}Sr/^{86}Sr$		Hodell et al. (1991)	
Sample Locality	Sample Number	Sampled Species	Measured Ratio	In-Run Uncertainty	Interval Number	Model Age
Subunit 1 Lithofacies 1 Oysters						
St. Michielsberg	2/15/89-V2	*Ostrea frons*	0.708804	0.000028	N-V	12.39
		duplicate	0.708809	0.000017	N-V	12.18
		average	0.708807		N-V	12.29
St. Michielsberg	2/15/89-V7	*Ostrea frons*	0.708784	0.000023	N-V	13.24
		duplicate	0.708814	0.000014	N-V	11.97
		average	0.708799		N-V	12.60
St. Michielsberg	2/15/89-V7	*Ostrea frons*	0.708830	0.000014	N-V	11.29
		duplicate	0.708833	0.000020	N-V	11.16
		average	0.708832		N-V	11.23
St. Michielsberg	2/15/89-V7	*Ostrea frons*	0.708781	0.000015	N-V	13.37
Average			0.708805		N-V	12.37
Modelled Hyperbolic Assymptote			0.708770		N-V	13.83
Subunit 2 Lithofacies 6 Oyster						
Seru Spreit	9/16/90-IIB	*Ostrea frons*	0.709008	0.000015	N-III	4.80
Subunit 2 Lithofacies 7 Oysters						
St. Annabaai	9/15/90-IA	*Ostrea frons*	0.709045	0.000014	N-I	2.25
St. Annabaai	9/15/90-IA	*Ostrea frons*	0.709059	0.000012	N-I	2.02
St. Annabaai	9/15/90-IA	*Ostrea frons*	0.709063	0.000011	N-I	1.95
Average			0.709056		N-I	2.07
Modelled Hyperbolic Assymptote			0.709030		N-I	2.50
Subunit 2 Lithofacies 7 Brachiopods						
St. Annabaai	9/15/90-B	*T. barretti*	0.709035	0.000019	N-1	2.42
St. Annabaai	9/15/90-B	*T. barretti*	0.709050	0.000017	N-1	2.17
St. Annabaai	9/15/90-B	*T. barretti*	0.709044	0.000018	N-1	2.27
St. Annabaai	9/15/90-B	*T. barretti*	0.709028	0.000008	N-1	2.54
		duplicate	0.709035	0.000019	N-1	2.42
		average	0.709032		N-1	2.48
Average			0.709039		N-I	2.35
Modelled Hyperbolic Assymptote			0.709030		N-I	2.50

TABLE 3.2. Sr isotope model ages of skeletal material from the Seroe Domi Formation. Skeletal $^{87}Sr/^{86}Sr$ values have been evaluated with the regression age model of Hodell *et al.* (1991).

Fig. 3.6), which is in agreement with field relationships and the *Praeorbulina glomerosa circularis* (Blow) range zone of 15.2 to 16.2 Ma (Berggren *et al.*, 1985; Fig. 3.2). Modelling also suggests that the replacement calcite is depleted in $\delta^{13}C$, Sr and Mg, enriched in Mn, and is similar in composition to surrounding calcite cements. An $^{87}Sr/^{86}Sr$ of 0.708832 in completely altered Lithofacies 1 *O. frons* shells implies that alteration took place no later than the Middle to Late Miocene transition at 11.2 $\pm$1.4 Ma (Table 3.2; Fig. 3.6).

A single specimen of *Ostrea frons* sampled from 8 m below the top of Subunit 2 Lithofacies 6 packstones at Seroe Spreit yields an $^{87}Sr/^{86}Sr$ of approximately 0.70901 ($\pm$0.00002). Although other shells were not analyzed from this horizon, covariations in $^{87}Sr/^{86}Sr$ versus Sr abundance is consistent with a low degree of meteoric alteration. This implies a maximum Late Miocene to Pliocene age of 4.8 Ma ±0.5 Ma (after Hodell *et al.*, 1991; Table 3.2; Fig. 3.6).

Subunit 2 Lithofacies 7 *T. barretti* and *O. frons* exhibit a difference of only 0.00001 between unaltered and completely altered shells. However, linear $^{87}Sr/^{86}Sr$ versus 1/Sr covariations and similarities between Subunit 1 and Subunit 2 textural alteration, suggest that mixing processes similar to those in Lithofacies 1 may also explain the Lithofacies 7 shell compositions. Unaltered Lithofacies 7 oysters and brachiopods have Sr isotope ratios of 0.70903 (±0.00002). This suggests an age of Late Pliocene age of 2.5 ±0.6 Ma, which is consistent with field relationships and biostratigraphy (Table 3.2; Fig. 3.6). Modelling suggests that the replacement calcites are depleted in $\delta^{13}C$ and Sr, and enriched in Mn. *O. frons* and *T. barretti* also exhibit a difference of 1‰ $\delta^{18}O$, which implies either differential initial compositions, extents of reaction, or meteoric fluid compositions. Completely replaced oyster and brachiopod skeletal material contains $^{87}Sr/^{86}Sr$ signatures of 0.709063 and 0.709050 respectively. This implies that shell alteration took place in meteoric fluids no later than the Late Pliocene to Pleistocene at 1.95 to 2.17 ±1.4 Ma soon after deposition of Lithofacies 7 grainstones.

Depositional History of the Seroe Domi Formation

The following is a chronologic synthesis of the multiple events of Seroe Domi Formation sedimentation and erosion that took place during tectono-eustatic changes in relative sea level. The overlying Quaternary Terrace Limestones (De Buisonjé, 1974; Bandoian and Murray, 1974; Herweijer and Focke, 1978; Schubert and Szabo, 1978; Stienstra, 1982) are also evaluated in order to provide estimates of the minimum age of the Seroe Domi Formation as well as post-depositional tectonism. A comprehensive description of the complex history of Cretaceous and Early Tertiary basement lithologies of Curaçao, on which the Seroe Domi Formation is deposited, is presented in Beets (1972, 1977), Beets *et al.* (1984) and Klaver (1987).

Subunit 1

The depositional age of Subunit 1 is suggested to be Middle Miocene based on the *Praeorbulina glomerosa* biostratigraphic range zone (Fig. 3.2) and the 13.8 +1.4 Ma Sr isotope model age (Fig. 3.7). Combined with the benthic foraminifer paleobathymetry, this suggests that Subunit 1 was deposited during the Middle Miocene as carbonate gravity flows interbedded with pelagic oozes in forereef and carbonate slope environments at water depths of at least 500 m (Fig. 3.7). The coral, algal, and bivalve skeletal fragments comprising Lithofacies 1, 3, and 4 sediments were generated in shallow-marine fringing reef environments. Therefore the isolated deep-water setting of Curaçao requires that the Curaçao basement acted as a platform on which carbonate production occurred. The strike and steep dip of Subunit 1 limestones, the small size and shape of the island, and the northwest-southeast normal faults surrounding Curaçao further suggest that Curaçao during the Middle Miocene was a relatively sheer-sided pedestal-like horst rising at least 500 meters above the surrounding sea floor. The deep-water deposition of Subunit 1 would therefore have taken place on the basal leeward flank of this submarine pedestal (Fig. 3.7). The mechanisms responsible for the observed restriction of Subunit 1 deposits to the southeastern half of the outcrop belt on Curaçao are uncertain.

No evidence exists as to whether Middle Miocene Curaçao had an emergent basement surrounded by fringing carbonate shelves similar to that of modern Curaçao, or whether it was a completely submerged shallow-water platform. However, sediment transport off the top of the Curaçao horst was probably most effective during sea level high-stands as it is in the modern Bahamas and St. Croix, where relative sea level and storm frequency are dominant controls on shallow to deep water transportation (Hine, 1983; Hubbard *et al.*, 1990). In addition, the leeward distribution of the Seroe Domi Formation is consistent with off-shelf storm and wave-related sediment transportation into fore-reef and carbonate slope environments (De Buisonjé, 1974; Hine, 1983; Hubbard *et al.*, 1985).

Off-bank sedimentation rates in the Bahamas that may be analogous to Seroe Domi Formation deposition are highly variable, and depend on the source area composition, platform geometry and trigger mechanism (Schlager and Cermack, 1984; Cook, 1983; Mullins, 1985). Single events are observed to deposit over 30 m of gravity flow sediments in days (Cook, 1983; Fulthorpe and Melillo, 1986) to thousands of years (Grammar and Ginsburg, 1992). Conversely, pelagic oozes in modern Bahamian carbonate slope environments analogous in composition to Lithofacies 2 wackestones are

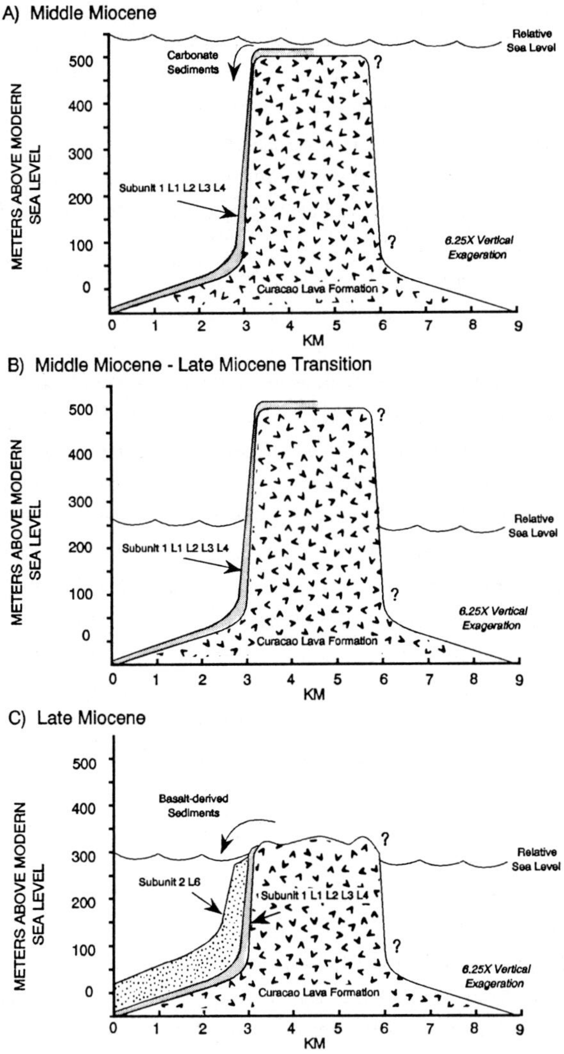

FIGURE 3.7. Schematic representation of the Mio-Pliocene depositional history of the Seroe Domi Formation on Curaçao. A. Middle Miocene deep-water deposition of Subunit 1 Lithofacies 1, 2, 3, and 4. B. Exposure of Subunit 1 lithologies to meteoric incursion due to the combined tectonic and eustatic lowering of relative sea level during the late Middle to early Late Miocene. C. Late Miocene erosion of the core of the island, initiating deposition of siliciclastic-rich Subunit 2 Lithofacies 6 lithologies.

deposited during extended periods of sedimentary quiescence associated with either high or lowstand sea level intervals (Cook, 1983). Therefore, the interbedding of grainstones and wackestones in Subunit 1 represents multiple gravity flow and quiescent sedimentation events of potentially very short duration. In addition, the absence of Lithofacies 2 wackestones in the upper half of the Subunit 1 sections suggests that Subunit 1 is either an upward shallowing sequence, or the result of seaward progradation of the fore-reef deposits (e.g. Mullins, 1985).

Subunit 2

The sedimentologic composition and stratigraphic succession of Subunit 2 represents an overall shallowing upward sequence from relatively deep water fore-reef environments to shallow *in situ* fringing reef environments. Subunit 2 Lithofacies 6 siliciclastic-rich packstones are consistent with the off-shelf transport of both carbonate and siliciclastic sediments into fore-reef and reef-front marine environments (Fig. 3.6). The siliciclastic sands in Lithofacies 6 packstones were derived from the Curaçao basement, and were presumably eroded in both subaerial and submarine environments. The stratigraphic positioning of Lithofacies 6 between Subunit 1 and Lithofacies 7, combined with the Sr isotope model age of 4.8 $\pm$1.4 Ma, suggests that Lithofacies 6 deposition took place during the Late Miocene to Early Pliocene (Fig. 3.6).

The coral assemblages observed in Lithofacies 5 suggest a Pliocene depositional age. Similarly, the coral biostratigraphy, Sr isotope model age of 2.5 $\pm$1.4 Ma, and cross-cutting stratigraphic relationships with Lithofacies 8 suggest that Lithofacies 7 is also Pliocene in age (Fig. 3.7). The *in situ* coral boundstones observed in both Lithofacies 5 and Lithofacies 7 are similar to reef-front deposits in St. Croix (Hubbard *et al.*, 1985), Panama (MacIntyre, 1983), Belize (James and Ginsburg, 1979) and Jamaica (Land and Moore, 1977), which suggests that they were deposited in shallow-water reef-front environments (Fig. 3.8). Reef-front sediment accumulation rates in the modern shelf margin on St. Croix are estimated to be 0.3 to 0.7 m/1000 years (Hubbard *et al.*, 1990). This implies that the respective 30 to 50 m-thick Lithofacies 5 and Lithofacies 7 sections could have been deposited within single 100,000 year-long third-order Neogene sea level highstands (Haq *et al.*, 1987; Greenlee and Moore, 1989).

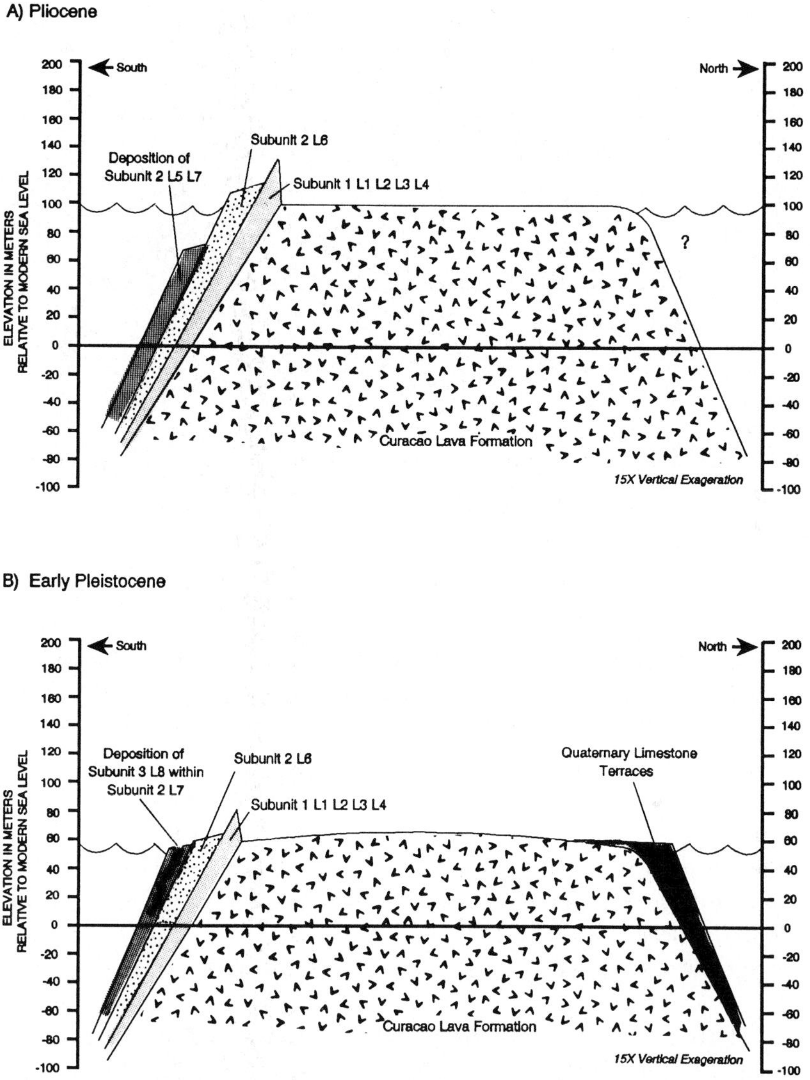

FIGURE 3.8. Schematic representation of the Plio-Pleistocene depositional history of the Seroe Domi Formation and Quaternary Limestone Terraces on Curaçao. A. Late Pliocene to Early Pleistocene erosion of Lithofacies 7, forming stratiform chutes, troughs and caves. B. Pleistocene filling of erosional cavities in Subunit 2 Lithofacies 7 with Subunit 3 Lithofacies 8 siliciclastic sandstones and conglomerates. The oldest of the Quaternary Limestone Terraces may have begun forming at this time.

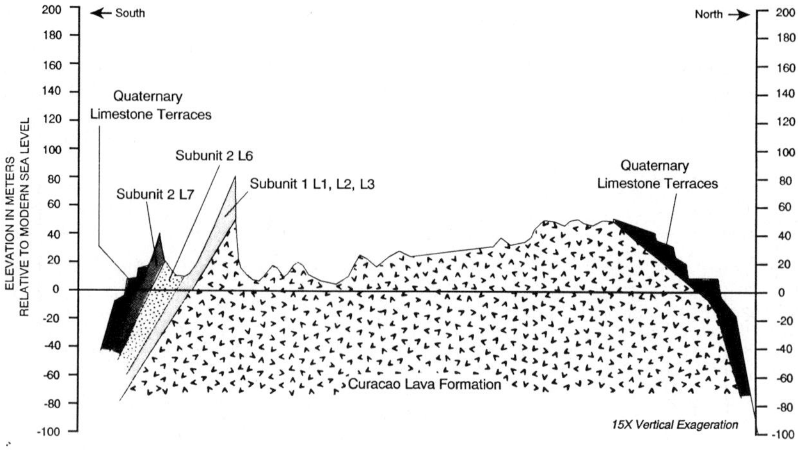

FIGURE 3.9. Geologic cross-section of present-day Curaçao. The cross-section extends from St. Michielsberg on the southwestern coast, to the Curaçao International Airport on the northeastern coast.

Subunit 3

Subunit 3 Lithofacies 8 contains basalt-derived sands and conglomerates, truncated horizontal bedding and coral fragments, which are suggestive of deposition in shallow marine shelf and reef-front environments (Fig. 3.8). And the presence of *Acropora palmata* corals suggests a maximum depositional age of Pleistocene. The oldest terrace directly onlapping Lithofacies 8 is the Middle Terrace I, which radiometric dating suggests is 0.5 Ma years old (De Buisonjé, 1974; Schubert and Szaro, 1984; Herweijer and Focke, 1984). Therefore, the depositional age of Lithofacies 8 is interpreted to be during the Early Pleistocene (Fig. 3.8).

Geologic History of the Quaternary Terrace Limestones

Approximately 25 percent of the emergent surface area of the island of Curaçao is composed of a series of flat-lying coralgal packstones, grainstones and boundstones called the Quaternary Terrace Limestones. Terrace deposition occurred on both the windward and leeward flanks of the island (Fig. 3.9; De Buisonjé, 1974), in lagoons in the center of the island, and in channels between topographically elevated Seroe Domi Formation mountains. The preserved remnants of the Quaternary Terrace Limestones within these channels are now in stratigraphic juxtaposition with the down-dip portions of the Seroe Domi Formation outcrops (Fig. 3.9).

De Buisonjé (1974) originally described 5 distinct horizons or steps within the Curaçao terrace sequence (Figs. 1.7 and 3.10), and interpreted them to represent *in situ* carbonate deposition in shallow-water shelf and reef-front environments during successive highstands of Pleistocene sea level. Presumed lithologic equivalents of these major terrace sequences were then described and mapped on La Orchila (Schubert and Valastro, 1976) and La Blanquilla (Schubert and Szabo, 1978). Ensuing studies by Bandoian and Murray (1974) and Herweijer and Focke (1978) on Curaçao, Bonaire and Aruba have determined that each of these stepped terrace horizons is actually a complex stacked sequence of two or more depositional and erosional events that are often laterally discontinuous (Fig. 3.10). In addition, at least four submerged terraces have been observed on Bonaire (Herweijer and Focke, 1978). Each of these studies attempted to estimate the tectonic uplift rate of the Aruba-LaBlanquilla island archipelago by applying either uranium-thorium or $\delta^{14}C$ dating techniques to fossilized corals and molluscs. The results of these studies yield low uplift rates that range from 0.02 to 0.08 m/1000 years, with an estimated average of 0.05 m/1000 years (Herweijer and Focke, 1978; Fig. 3.10).

Relative Sea Level History of Curaçao

The above sedimentologic, biostratigraphic and geochemical analyses of the Seroe Domi Formation suggest that the Middle Miocene to Quaternary was a time period of extremely variable elevations in relative sea level that were driven by the interplay of both tectonic and eustatic forces. In the following discussion, the magnitude and relative impact of tectonism and eustasy will be evaluated. Uncertainties in age and depositional water depth have necessitated that the following reconstructions of relative sea level change are bracketed with high and low estimates based on the minimum and maximum age and paleobathymetic interpretations of the formation. Global sea level reconstructions suggest that the Middle Miocene experienced four individual third-order sea level highstands during this time period (Haq *et al.*,

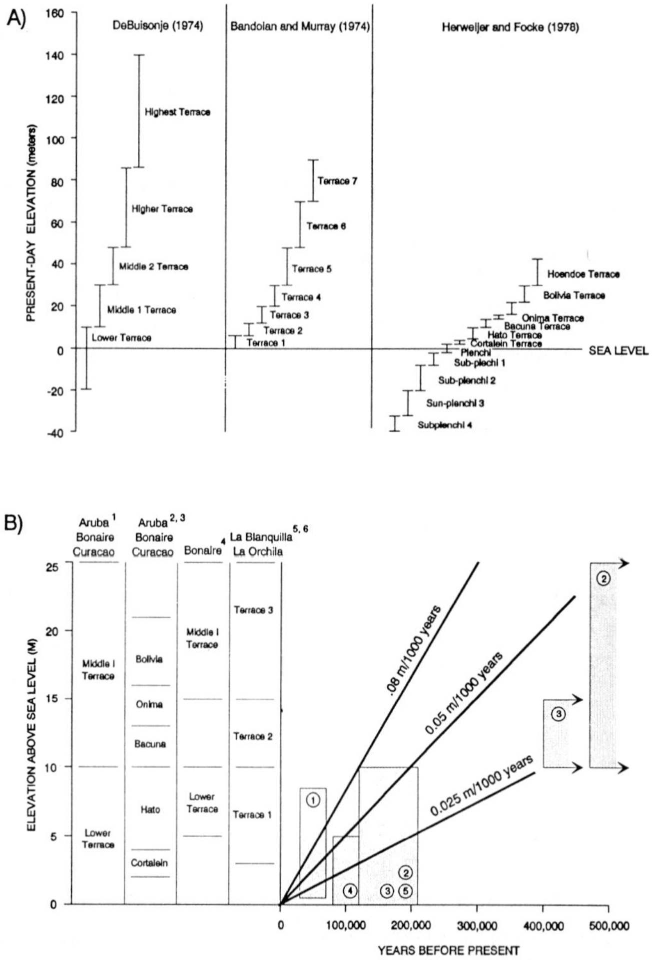

FIGURE 3.10. Summary and evaluation of previous analyses of the Quaternary Limestone terraces on Curaçao, Aruba, and Bonaire. A. Discrepancies between successive studies of the stratigraphy and nomenclature of the Quaternary Limestone Terraces on Curaçao and Bonaire, Netherlands Antilles. B. Cross-plot of the radionuclide age determined for samples from analyses of the Quaternary Limestone Terraces listed in Table 6, versus the stratigraphic nomenclature and range in sampling elevation listed for each sample. These uncertainties in stratigraphic control and age suggest that uplift rates during the Quaternary may have ranged from 0.025 to over 0.08 m/1000 years, which Herweijer and Focke (1978) suggest yields an average uplift rate of 0.05 m/1000 years. References: 1 = De Buisonjé (1974); 2 = Herweijer and Focke (1978); 3 = Schubert and Szabo (1978); 4 = Bandoian and Murray (1974); 5 = Schubert and Valastro (1976).

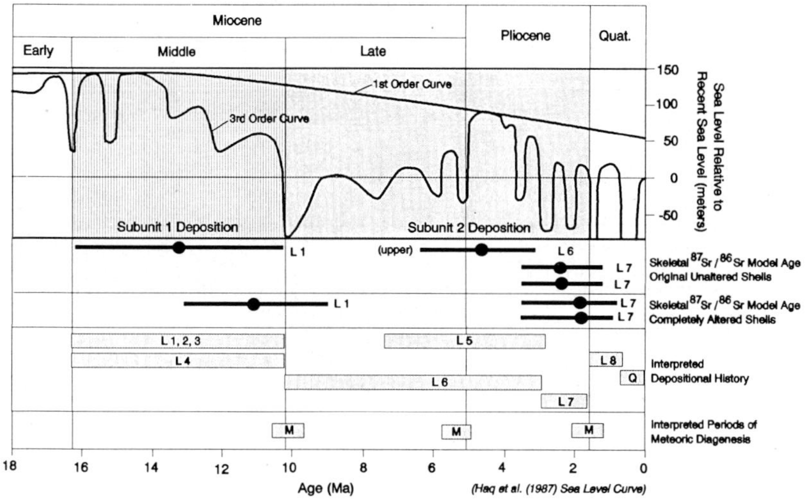

FIGURE 3.11. Chronostratigraphic summary of the Seroe Domi Formation.

1987). Unfortunately, the direct correlation of Subunit 1 deposition with any one these individual sea level highstands is not possible due to coarse chronologic resolution (Fig. 3.11) and discrepancies in the magnitude and periodicity of sea level change between published curves (Haq *et al.*, 1987; Greenlee and Moore, 1989). In addition, the platform source areas of Subunit 1 sediments have been completely eroded, preventing the distinction of aggradational from progradational sequences.

Subunit 1 Lithofacies 1, 2, and 3 deep-water sedimentary gravity flows and pelagic oozes are in stratigraphic juxtaposition with Subunit 2 Lithofacies 7 shallow-water reef-front deposits at St. Michielsberg and Seru Blandan (Fig. 3.8). This demonstrates that the Curaçao horst has experienced hundreds of meters of vertical tectonic uplift and erosion during the Middle to Late Miocene and Pliocene. The elevated sequence of Quaternary Limestone Terraces on the modern island of Curaçao suggest that this tectonic uplift continued into the Pleistocene and Holocene (De Buisonjé, 1974; Herweijer and Focke, 1978). Extrapolation from the minimum 500 m water depth of Subunit 1 deposition suggested by benthic foraminifers in Subunit 1 Lithofacies 2 wackestones implies that Curaçao horst has experienced an average tectonic uplift rate of 0.03 m/1000 years since the Middle Miocene (Figs. 3.11 and 3.12).

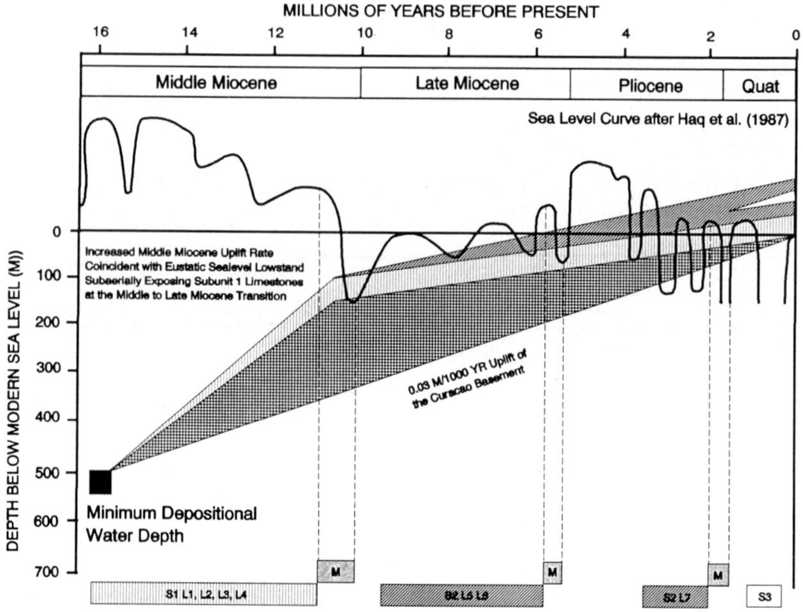

FIGURE 3.12. Interpreted depositional and diagenetic history of the Seroe Domi Formation on Curaçao. This diagram illustrates the relative timing of Seroe Domi Formation deposition and limestone diagenesis in relation to uplift of the Curaçao horst and third-order changes in global sealevel observed by Haq *et al.* (1987). The depositional sequences are represented by the following: Subunit I Lithofacies 1, 2, 3, and 4 = S1 L1 L2 L3 L4; Subunit 2 Lithofacies 5 and 6 = S2 L5 L6; Subunit 2 Lithofacies 7 = S1 L7; and Subunit 3 Lithofacies 8 = S3 L8. Each episode of meteoric diagenesis (M) correlates with a significant drop in third-order global sealevel. The overall uplift rate of 0.03 m/1000 yr was probably increased during the Middle Miocene to rates of at least 0.05 m/1000 years, which acted in conjunction with a eustatic drawdown to subaerially expose Subunit 1 limestones during the late Middle Miocene to early Late Miocene.

The Sr isotope model age of 11.2 ±1.4 Ma for completely altered Subunit 1 oyster shells suggests that Subunit 1 limestones may have been subaerially exposed as early as the late Middle Miocene or early Late Miocene (Figs. 3.11 and 3.12). This age correlates with global third-order sea level draw downs of approximately −200 m associated with the onset of Antarctic glaciation observed by Haq *et al.* (1987). The 500 m minimum depositional water depth interpreted for Subunit 1, in combination with this eustatic lowstand, suggests that the Curaçao horst would have required tectonic uplift rates of at least 0.05 m/1000 years during the Middle Miocene in order to subaerially expose Subunit 1 by the Middle to Late Miocene transition (Fig. 3.12). This is slightly higher than the estimated average Neogene uplift rate of 0.03 m/1000 years, and suggests that the Curaçao horst has experienced variable rates of uplift during the Mio-Plio-Pleistocene.

The Pliocene and Early Pleistocene shallow shelf to reef-front depositional environments of Subunit 2 Lithofacies 5 and 7 and Subunit 3 Lithofacies 8 suggest that the uplift rates of approximately 0.05 m/1000 years interpreted for the Quaternary Terrace Limestones (Herweijer and Focke, 1978; Schubert and Szabo, 1978) are also reasonable estimates of tectonic uplift during the Pliocene and Early Pleistocene (Figs. 3.11 and 3.12). However, the less than 100 m water depths associated with Lithofacies 5, 7, and 8 deposition could have been attained during any of the multiple eustatic highstands during the Plio-Pleistocene. However, these Plio-Pleistocene uplift rates are consistent with the average Neogene uplift rate of 0.03 m/1000 years estimated from Subunit 1 deposition (Fig. 3.12).

The Middle Miocene to Quaternary uplift observed in the Seroe Domi Formation is consistent with interpretations of the regional tectonic evolution of the Venezuela borderland. Muessig (1984) suggests that deformation caused by the oblique collision of the Caribbean and South American plates has largely been accommodated by folding and thrusting at the Southern Caribbean Deformed Belt accretionary wedge, and motion along northwest trending normal faults such as those surrounding Curaçao. Evidence from the present study of the Seroe Domi Formation suggests that tectonic uplift was initiated in the Middle Miocene, bringing the upper surface of the Curaçao basement into shallow marine photic zone environments, which triggered carbonate production and the resulting deep-water deposition of Subunit 1. These compressional forces on the Bonaire Block and the associated tectonic uplift of the Curaçao horst has then continued at consistently low rates into the Pleistocene and Holocene as evidenced by the Quaternary Limestone Terraces (Herweijer and Focke, 1978).

Conclusions

Subunits 1, 2, and 3 of the Seroe Domi Formation are an upward shallowing sequence of mixed carbonates and siliciclastics discontinuously deposited on the flanks of the Curaçao horst during the Miocene, Pliocene, and Pleistocene. The presence of *Praeorbulina glomerosa circularis* (Blow) and the absence of *Orbulina* suggests that Subunit 1 Lithofacies 1, 2, 3, and 4 were deposited during the early Middle Miocene at 15.2 to 16.2 Ma. A $^{87}Sr/^{86}Sr$ model age from oyster shells suggest a corroborative age of 13.8 ±1.4 Ma. Subunit 2 Lithofacies 5 coral biostratigraphy suggests a depositional age of Late Miocene to Early Pliocene. Sr isotope regression age modeling of Subunit 2 Lithofacies 6 oysters implies a Late Miocene to Early Pliocene age of 4.8 ±1.4 Ma. Oyster and brachiopod $^{87}Sr/^{86}Sr$ in Subunit 2 Lithofacies 7 grainstones yields a Pliocene age of 2.5 ±1.4 Ma, which is also consistent with stratigraphic field relationships. Subunit 3 Lithofacies 8 coral biostratigraphy and the stratigraphic onlapping of Late Pleistocene reef terraces suggests a Lithofacies 8 depositional age of Early Pleistocene.

The island of Curaçao is a fault-bounded horst that has experienced hundreds of meters of tectonic uplift during the Middle and Late Miocene, as demonstrated by the stratigraphic juxtaposition of deep-water Subunit 1 sediment gravity flows and pelagic oozes with shallow-water Subunit 2 coralgal boundstones. Subunit 1 Lithofacies 1, 2, 3, and 4 were deposited during the Middle Miocene near the base of this relatively shear-sided submarine horst. Geochemical evidence from skeletal material suggests that Subunit 1 limestones were stabilized in meteoric fluids as early as the late Middle Miocene, suggesting that tectonic uplift rates of approximately 0.03 m/1000 years combined with third-order eustatic lowstands to expose the island during the Late to Middle Miocene transition. This triggered intensive weathering and erosion of the basement, which initiated Subunit 2 Lithofacies 6 siliciclastic deposition in fore-reef to reef-front environments. The deposition of Late Miocene to Early Pliocene Lithofacies 5 and late Pliocene Lithofacies 7 reef-front boundstones suggest that post-Middle Miocene tectonic uplift rates were consistently low at approximately 0.05 m/1000 years, and have remained at these relatively quiescent rates through the Pleistocene and into the modern. Shell geochemistry and petrography suggest that Lithofacies 7 deposits were exposed to meteoric alteration as early as the Late Pliocene. Subunit 3 Lithofacies 8 siliciclastics were then deposited within stratiform chutes and caves in Lithofacies 7 in the Early Pleistocene. Quaternary Terrace Limestones have since been deposited in the Late Pleistocene, onlapping Seroe Domi Formation Subunit 2 and 3 deposits.

CHAPTER 4

DOLOMITIZATION

In Chapters 2 and 3 a sedimentologic, stratigraphic and chronologic framework was established for the Seroe Domi Formation limestones on Curaçao (Fig. 4.1). Three different types of dolomite have been observed within the Seroe Domi Formation, which are herein called Dolomites I, I' and II. Each of these dolomites have distinct geographic and stratigraphic distributions, petrographic characteristics, and geochemical compositions. In this chapter, Dolomites I, I', and II will be evaluated with respect to the geological framework in which they occur, in order to: 1) reconstruct the composition, timing, chemical selectivity, and paleohydrology of the fluids that dolomitized the limestones; 2) document the intermediate stages of the dolomitization process across dolomite-to-limestone transition zones; and 3) to correlate these diagenetic events with respect to relative sea level history. To accomplish this, several independent lines of geologic evidence have been intergrated, including the: 1) petrography and distribution of the dolomites; 2) diagenetic sequence of the host limestones; and 3) geochemical analyses of the dolomites and the entire paragenetic sequence; 4) evaluation of modern groundwater chemistries.

Previous Studies of Seroe Domi Formation Dolomitization

No previous detailed petrographic or geochemical analyses of the Seroe Domi Formation dolomites on the island of Curaçao have previously been completed. However, several studies have addressed the dolomitization of the Seroe Domi Formation limestones on Bonaire. Deffeyes *et al.* (1965) observed that dolomites occur in the up-dip portions of the Seroe Domi Formation limestones in northern Bonaire, and interpreted them to have precipi-

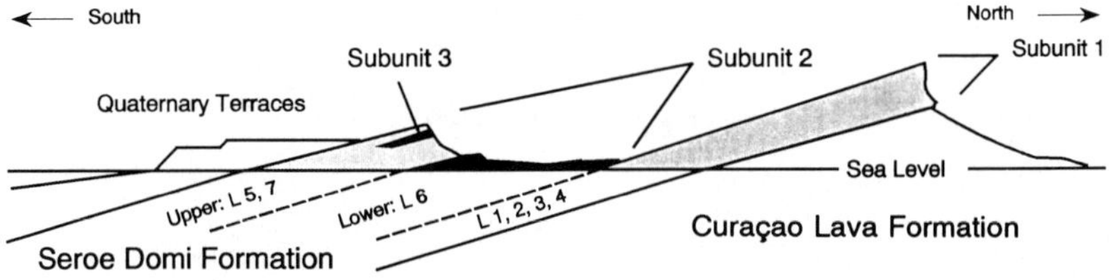

Subunit 1: Middle Miocene Interbedded Coralgal Grainstones, Foraminifer Wackestones, and Allochthonous Blocks
Lithofacies 1, 2, 3, and 4 containing Dolomite I and I'
Oyster $87Sr/86Sr$ = 0.708781 - 0.708832
Pelagic Foraminifer *Praeorbulina glomerosa circularis* Blow
Benthic Foraminifer *Melonis pompiloides* Fitchell and Moll
Lower Subunit 2: Late Miocene to Pliocene Siliciclastic Packstones and Grainstones
Lithofacies 6 exhibits no dolomitization
Oyster $87Sr/86Sr$ = 0.709008
Upper Subunit 2: Late Miocene to Pliocene to Early Pleistocene Coralgal Grainstones and Boundstones
Lithofacies 5 containing Dolomite I'
Lithofacies 7 containing Dolomite II
Brachiopod and Oyster $87Sr/86Sr$ = 0.709045 - 0.709063
Scleractinian Biostratigraphic Range Overlap
Subunit 3: Early Pleistocene Siliciclastic Sandstones and Grainstones
Lithofacies 8 exhibits no dolomitization
Deposited in Eroded Upper Subunit 2 Beds
Scleractinian *Acropora palmata*

FIGURE 4.1. Geologic cross-section summarizing the Curaçao Lava Formation, Seroe Domi Formation, and Quaternary Limestone Terraces succession on Curaçao. Cross-section is oriented perpendicular to the leeward coast of Curaçao at St. Michielsberg. Listed below the cross-section is a summary of the sedimentologic, chronostratigraphic, and diagenetic evidence observed in Subunits 1, 2, and 3 of the Seroe Domi Formation.

tated from gravity-driven hypersaline brines derived from the evaporation of seawater. This implies that the most intense dolomitization of limestone is concentrated in up-dip regions of the lithology proximal to the source of the evaporitic fluids (Deffeyes *et al.*, 1965). Hypersaline fluids and dolomitized muds from a modern salina on Bonaire called the Pekelmeer were used by Deffeyes *et al.* (1965) as a modern analogue for the Seroe Domi Formation dolomites. However, the hydrologic applicability of the Pekelmeer was later drawn into question when Lucia (1968) and Murray (1969) found that the proper density stratification for gravity-driven flow is attained only periodically throughout the year due to climatic and tidal variations in brine salinity.

Controversy developed over whether the dolomitizing fluids were composed of hypersaline brines or mixtures of seawater and freshwater. Sibley (1980, 1982) documented that dolomitization of the Bonaire limestones had taken place after subaerial exposure. This is based on the observation that bladed calcite cements encrusted by dolomite cements were precipitated within leached skeletal molds. This petrographic evidence, combined with positive $\delta^{18}O$ (+2 to +4‰ PDB) and low Na concentrations (280 ppm) in the dolomite, caused Sibley (1980) to interpret the dolomites as having been precipitated from the mixing of freshwater and seawater. Conversely, a recent petrographic and geochemical study by Lucia and Major (in press) again concludes that the Bonaire dolomites were precipitated from hypersaline brines. This is based on: 1) up-dip stratiform dolomite distributions in the limestones; 2) positive $\delta^{18}O$ (+0.5 to +2‰ PDB) in the dolomites, which are similar to Recent Pekelmeer dolomites (Major *et al.*, 1992); and 3) the presence of anhydrite inclusions within the dolomites.

The Seroe Domi Formation dolomites on Curaçao were first mentioned by De Buisonjé (1974), who observed that his Older Seroe Domi stratigraphic unit (which is approximately equivalent to Subunit 1 Lithofacies 4 in this study) was massively dolomitized. In addition, he observed stratiform to pervasive up-dip dolomite lenses in his Middle and Younger Seroe Domi units (which is approximately equivalent to Subunit 2 Lithofacies 6 and 7 in this study). De Buisonjé (1974) also recorded the presence of dolomitized limestone surrounding a fracture at Veerisberg and dolomitized allochthonous blocks at Blauwbaai. However, although dolomite was observed at these localities, neither of these distinct dolomite distributions were observed in the present study. Seaward-dipping dolomite bodies are also reported from drill cuttings and outcrops in the Bullenbaai region on Curaçao by O. Meyer (unpublished manuscript) in a report to assess the feasibility of commercially mining limestone. And finally, two additional reports of Seroe Domi Formation dolomitization resulted from studies of limestone phosphatization at the Tafelberg mountain in southeastern Curaçao by Ten Have *et al.* (1982) and Stienstra (1991). Petrographic relationships illustrated in these studies indicate that dolomite was precipitated prior to phosphatization, then extensively leached and replaced by phosphate cements.

Diagnostic Petrographic Criteria for Dolomites I, I' and II

Three types of dolomite with distinct petrographic characteristics have been observed within the Seroe Domi Formation, which are designated as Dolomite I, I', and II.

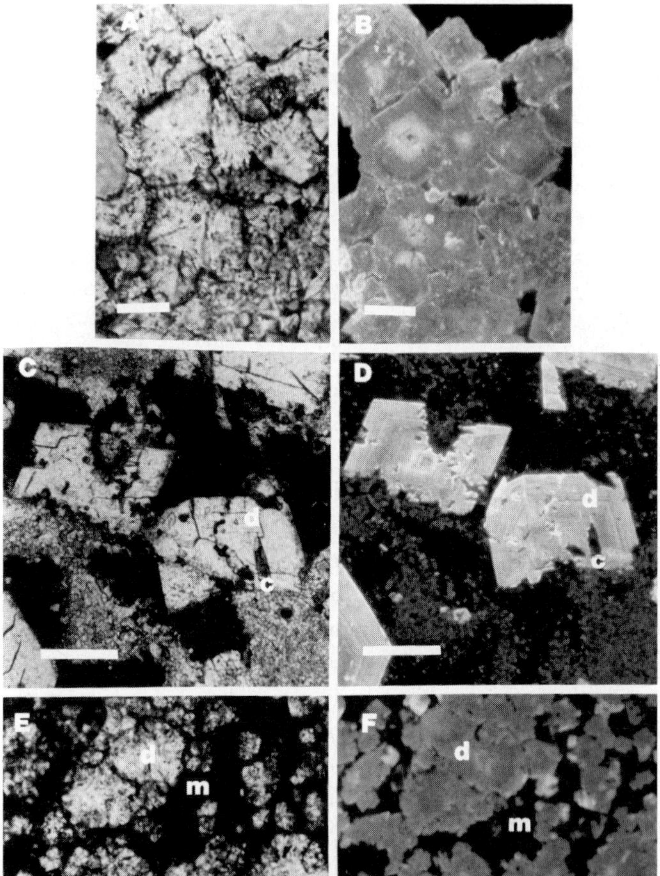

FIGURE 4.2. Petrography of Dolomite I. A and B. Paired plane-light and CL photomicrographs of subhedral to euhedral Dolomite I rhombohedra in Subunit 1 Lithofacies 1 grainstones at Seru Blandan. Sample number 3/29/89-II4. Scale bar is 100 μm. C and D. Paired plane-light and CL photomicrographs of euhedral Dolomite I rhombohedra (d) encrusting pre-dolomite bladed calcite cements (C) in Subunit 1 Lithofacies 1 grainstones at St. Michielsberg. Sample number 2/14/89-II5. Scale bar is 100 μm. E and F. Paired plane-light and CL photomicrographs of anhedral Dolomite I crystals replacing micrite in Subunit 1 Lithofacies 2 wackestones at Seru Blandan. Sample number 3/29/89-II1. Scale bar is 50 μm.

Dolomite I and I' Petrography

Dolomites I and I' are composed of large rhombohedral crystals that are 75 to 300 μm in diameter. Crystal textures include both planar and nonplanar euhedral to anhedral dolomite, all of which exhibit unit extinction (Figs. 4.2 and 4.3). The primary petrographic difference between Dolomite I and I' is in their cathodoluminescent (CL) concentric crystal zoning. Dolomite I crystals exhibit a bright yellow-orange CL with fine-scale hairline concentric zoning throughout the crystals, with each concentric zonation ranging from less than 1 μm to 10 μm in thickness (Fig. 4.2). Conversely, Dolomite I' crystals exhibit a bright orange-brown CL with wide dark concentric zones up to 30 μm in thickness (Fig. 4.3). All Dolomite I' crystals are terminated by a 5 m-thick extinct outer rim (Fig. 4.3). Dolomite I' crystals also exhibit sectoral zoning in CL, which is petrographically described and geochemically documented in Fouke and Reeder (1992).

Fluid inclusions are extremely rare in Dolomites I and I', and composed of single-phase inclusions less than 3 μm in diameter (observations by A. Cox). Conversely, elongate to equidimensional solid-phase calcite inclusions 3 to 25 μm in length are common in Dolomite I and I'. The concentric CL zoning in both Dolomite I and I' continues uninterrupted around large and small inclusions (Figs. 4.2 and 4.3). Four distributions of these calcite inclusions have been observed, including: 1) columnar calcite cements growing on the surface of skeletal grains and encrusted by dolomite; 2) linear to curvilinear arrays of inclusions oriented parallel to one another and radiating off of a common surface; 3) randomly distributed and randomly oriented inclusions throughout the host dolomite rhombohedra; and 4) inclusions concentrated in the middle of crystals, creating rhombohedra with cloudy centers and clear rims (CCCR of Sibley, 1980; Figs. 4.2 and 4.3). Some of these calcite inclusions have been selectively dissolved, leaving elongate crystal molds within the dolomite crystals that are similar to those observed in Seroe Domi Formation dolomites on Bonaire (Sibley, 1980).

Dolomite II Petrography

Dolomite II is distinct in crystal size and CL zoning patterns with respect to Dolomites I and I'. Dolomite II crystals are relatively small, measuring only 50 to 75 μm in apparent diameter compared to the 75 to 300 μm diameter

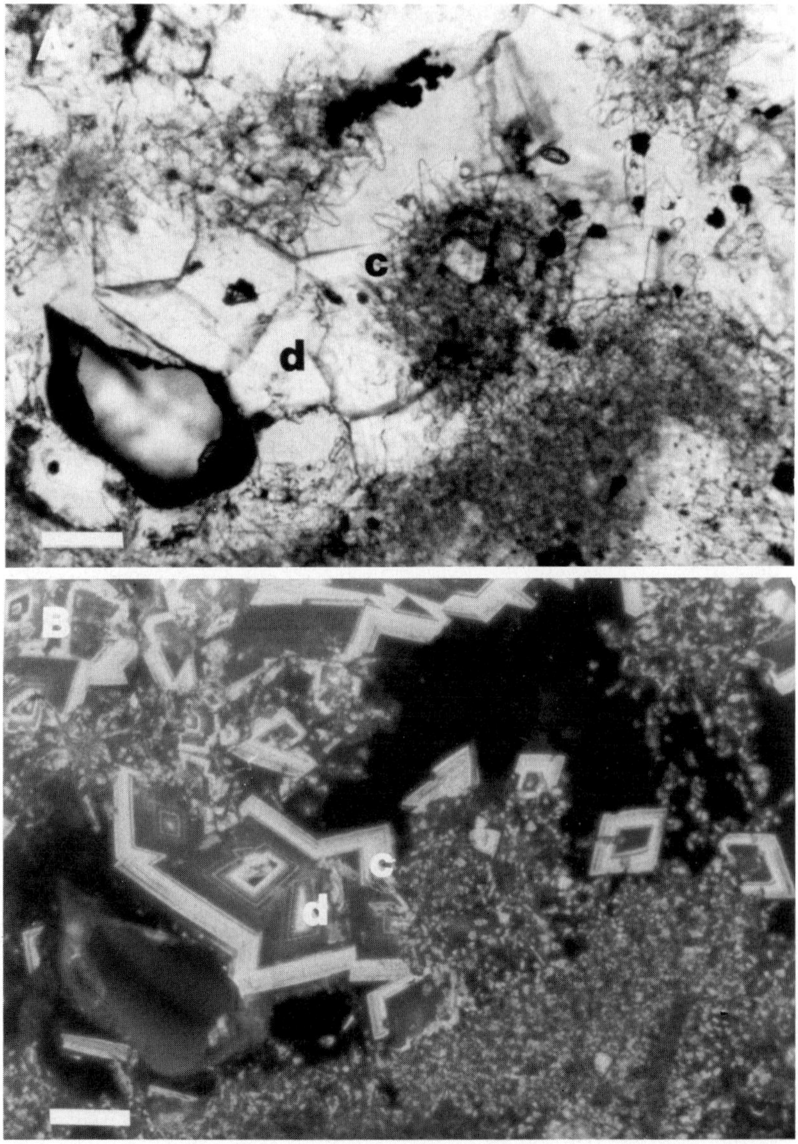

FIGURE 4.3. Petrography of Dolomite I'. A and B. Paired plane-light and CL photomicrographs of euhedral Dolomite I' rhombohedra (d) encrusting pre-dolomite bladed calcite cements (c) in Subunit 1 Lithofacies 1 grainstones at Kabrietenberg. Sample number 6/21/88-V3. Scale bar is 100 μm. C and D. Paired plane-light and CL photo-

micrographs of euhedral Dolomite I' rhombohedra (d) with inclusions of pre-dolomite bladed calcite cements (c) in Subunit 1 Lithofacies 1 grainstones at Seru Magasina. Sample number 7/2/88-III1. Scale bar is 100 μm.

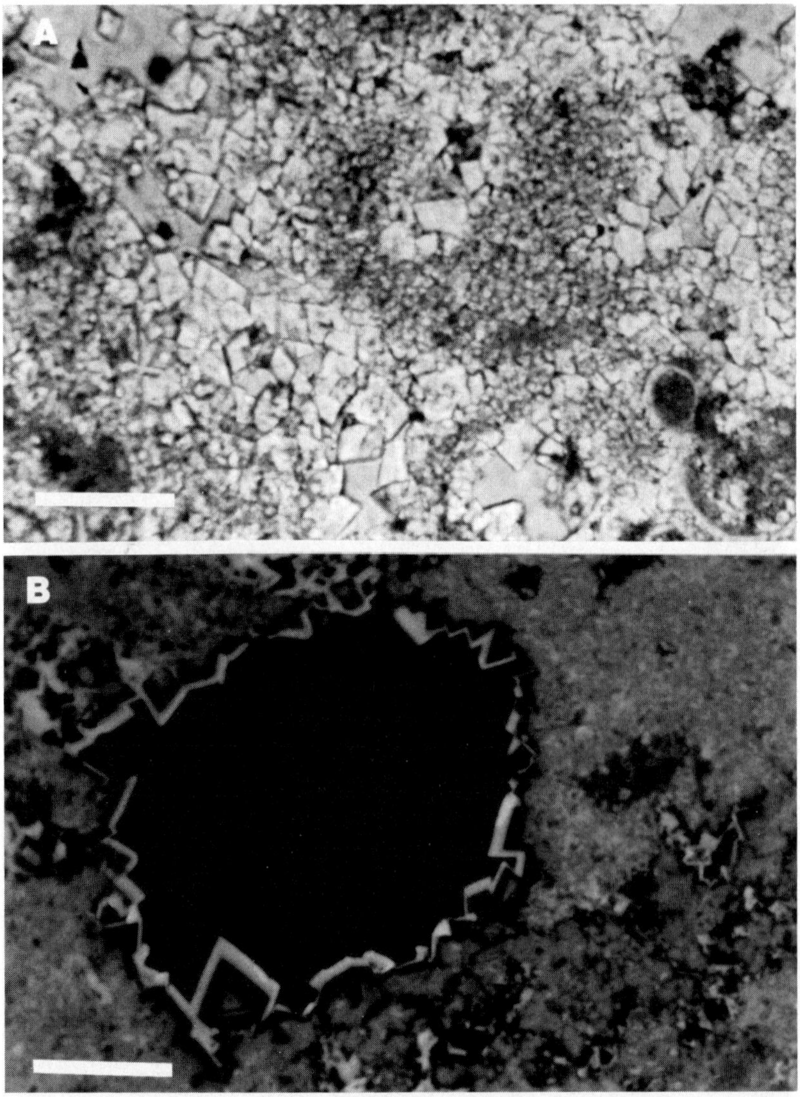

FIGURE 4.4. Petrography of Dolomite II. A. Plane-light photomicrograph of euhedral Dolomite II rhombohedra in Subunit 2 Lithofacies 7 grainstones at Seru Spreit. Sample number 9/16/90-IIa. Scale bar is 75 μm. B. CL photomicrograph of euhedral Dolomite II rhombohedra in Subunit 2 Lithofacies 7 grainstones at Seru Spreit. Sample number 9/16/90-IIa. Scale bar is 75 μm.

crystals of Dolomites I and I'. Dolomite II crystals exhibit a mixture of sucrosic planar euhedral to subhedral dolomite textures (Fig. 4.4). However, nonplanar anhedral dolomite microspar such as that observed in Dolomite I and I' is not found in Dolomite II.

Dolomite II also has unique concentric CL crystal zoning colors and patterns, with a dull CL core up to 20 μm in diameter, rimmed by an extinct CL zone 10 μm-thick, followed by a bright orange CL outer concentric zone 10 to 20 μm in thickness with fine-scale hairline zoning (Fig. 4.4). The extinct CL middle zones are only observed in crystals growing into void space, while the matrix crystals are generally composed of a mixture of the dull core CL and the bright outer zone CL (Fig. 4.4). Neither fluid inclusions or elongate solid-phase calcite inclusions that occur in Dolomite II were observed in Dolomite I and I' (Fig. 4.4).

Distribution and Correlation of Dolomites I, I', and II

Dolomites I, I', and II have distinct geographic and stratigraphic distributions within the Seroe Domi Formation outcrop belt, and never co-occur in the same lithofacies at any one locality (Table 4.1; Fig. 4.5). Limestones containing either Dolomite I, I', or II exhibit varying extents of dolomitization that have been separated into the following categories: 1) *non-dolomitized* limestone that contains no dolomite (0% dolomite); 2) *partially dolomitized* limestones containing 1 to 75% dolomite, with variable proportions of cement versus replacement dolomite; and 3) *pervasively dolomitized* limestones containing 75 to 100% dolomite, with variable proportions of cement versus replacement dolomite. In addition, the pervasively and partially dolomitized limestone bodies observed in the Seroe Domi Formation exhibit two general types of geometries, which include: 1) *stratiform* dolomite bodies that parallel the original planar bedding geometries of the precursor limestones; and 2) *bed cross-cutting* dolomite bodies that cut across the original planar bedding geometries of the precursor limestones. The terms *massive* and *pervasive* dolomitization are often used synonymously. However, pervasive dolomitization in the Seroe Domi Formation occurs as planar stratiform bodies as well as massive bed cross-cutting bodies. Therefore, in order to avoid incorrect implications regarding geometry of the dolomite bodies, the term *pervasive* will be used to describe completely dolomitized limestones.

LOCALITY NAME	SUBUNIT	LITHOFACIES	DOLOMITE TYPE	DOLOMITE DISTRIBUTION
1) Seru Belefa	2	7	II	L
2) Knipberg	2	7	II	L
3) Seru Zjeremi	2, 3	7, 8	II	L
4) Ser'i Baha So	2	7	II	L
5) Ser'i Kalki	2	7	II	L
6) Seru Bo Kuki	2	7	II	L
7) Ser'i Boka	2	7	II	L
8) Seru Grandi	2	7	II	L
9) Seru Largu	2	7	II	L
10) Seru Lakete	2	7	II	L
11) Seru We'i Awa	2	7	II	L
12) Seru Nobo	2	7	II	L
13) Seru Kabaye	2, 3	7, 8	II	L, S
14) Seru Kabritu	2	7	II	L
15) Seru Rinkon	2	7	II	L
16) Seru Lucia	2	7	II	L
17) Seru Largu	2	7	II	L
18) Seru Popchi	2	7	II	L
19) Seru Hermanus	2	7	II	L, S
20) Ser'i Mahoc	2	7	II	L
21) Seru Pais	2	7	II	L
22) Ser'i Kabritu	1, 2	1, 6, 7	I	S
23) Seru Sami	1, 2	1, 6, 7	I, II	P
24) Seru Spreit	1, 2	1, 6, 7	I, II	S
25) Sint Michielsberg	1, 2	1, 2, 3, 6, 7	I, II	S
26) Seru Blandan	1, 2	1, 2, 3, 6, 7	I, II	S
27) Blauwbaai	2	7	II	L
28) Jack Evertszberg	1, 2	1, 6, 7	I	S
29) Veerisberg	1, 2	1, 6, 7	I	S
30) Seru Pretu	1, 2	1, 6, 7	I	L
31) Planters Rust	2, 3	7, 8	II	L
32) Ser'i Domi	1, 2	1, 3, 6, 7	I	S
33) Sint Annabaai	2, 3	6, 7, 8	II	L
34) Fort Nassau	1, 2, 3	1, 6, 7, 8	I	S
35) Ser'i Salina Abou	2	7	I	S
36) Ser'i Kura	1, 2	1, 6, 7	I	L
37) Seru Spano	2	7	I	L
38) Seru Mansinga	1	5	I'	P
39) Kabrietenberg	1, 2	1, 4, 6, 7	I'	P
40) Ser'i Boka	1, 2	1, 4, 6, 7	I'	P
41) Tafelberg	1?	1?, 4, +?	I'	P
42) Seru Magasina	1	1, 4	I'	P
43) Seru Prekstul	1	1, 4	I'	S, P
44) Duivelsklip	1	4	I'	P

Dolomitization Distributions:
S = stratiform; P = pervasive bed cross-cutting; L = localized
Dolomite Types:
I = Dolomite I; I' = Dolomite I', II = Dolomite II

TABLE 4.1. Locality names, lithofacies, types of dolomite, and dominant dolomitization fabrics of individual flatiron exposures of the Seroe Domi Formation on Curaçao. Numbers refer localities in Figure 4.5.

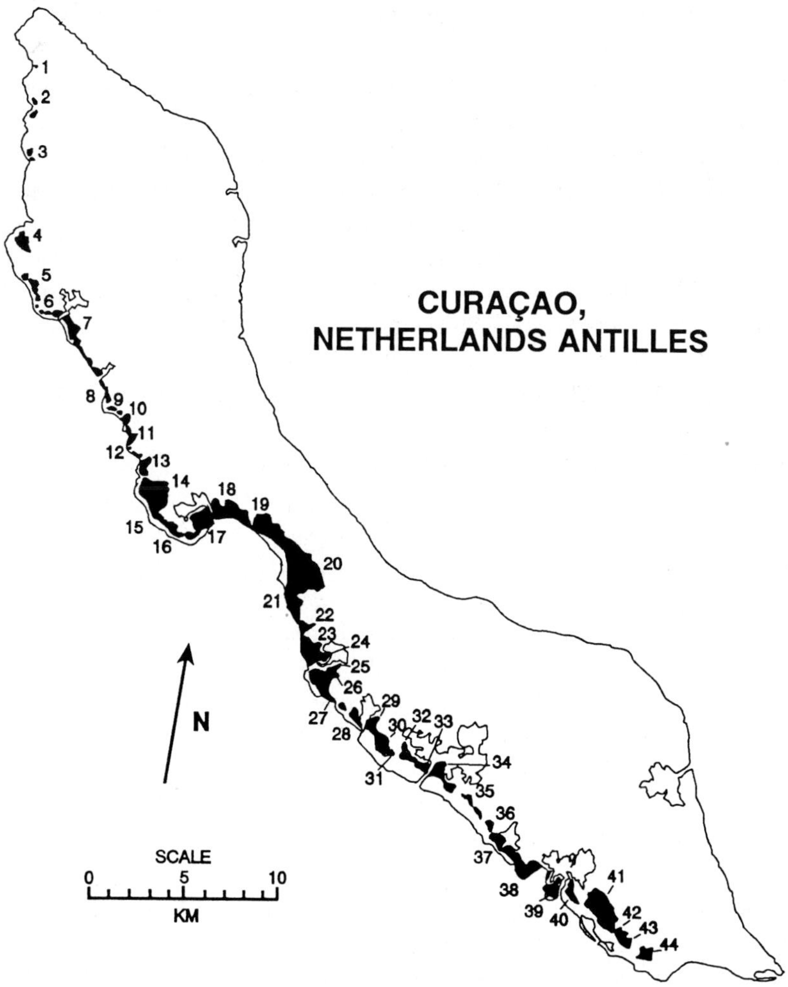

FIGURE 4.5. Geographic map of the island of Curaçao, Netherlands Antilles. The Seroe Domi Formation outcrop belt on the leeward coast of Curaçao is in black (after De Buisonjé, 1974). Numbered outcrops refer to localities and dolomite distributions presented in Table 4.1.

Geographic Distribution

The distributions of Dolomite I, I', and II have been geographically mapped throughout the Seroe Domi Formation outcrop belt on Curaçao, as well as stratigraphically mapped within each individual mountain (Table 4.1; Figs. 4.5 and 4.6; Fouke, 1993). Dolomite I has been observed exclusively in the northwestern portion of Middle Miocene Subunit 1 Lithofacies 1, 2, and 3 deposits that crop out from Ser'i Kabritu to Seru Kura (Localities 22 to 36 in Fig. 4.5; Fig. 4.6). Conversely, Dolomite I' occurs within the southeastern exposures of Middle Miocene Subunit 1 Lithofacies 1, 2, and 4 deposits that crop out from Kabrietenberg to Duivelsklip (Localities 39 to 44 in Fig. 4.5; Fig. 4.6). Dolomite I' also occurs in Late Miocene-Early Pliocene Subunit 2 Lithofacies 5 deposits at Seru Magasina (Locality 38 in Fig. 4.5; Fig. 4.6). Dolomite I' is the only Seroe Domi Formation dolomite observed within two different subunits. And finally, Dolomite II occurs exclusively within Late Pliocene-Early Pleistocene Subunit 2 Lithofacies 7 deposits, which extend northwestward across the length of the leeward coast of Curaçao from Seru Spano to Seru Belefa (Localities 1 to 37 in Fig. 4.5; Fig. 4.6). The Late Miocene Subunit 2 Lithofacies 6 siliciclastic-rich coralgal grainstones, and Early Pleistocene Subunit 3 Lithofacies 8 siliciclastic sandstones are the only lithologies of the Seroe Domi Formation that do not contain dolomite.

The presence of dolomite within the Quaternary Limestone Terraces would be extremely important in determining the timing of dolomitization events within the Seroe Domi Formation. However, no dolomite was observed within the Quaternary Limestone Terraces on Curaçao during the present study. Stienstra (1991) reports small localized pockets of dolomite at the Tafelberg quarry within breccias that he interprets as Quaternary Limestone Terraces deposits based on mapping by De Buisonjé (1974). However, the patchy occurrence and fabric-destructive diagenesis of these breccias makes their stratigraphic and chronologic correlation difficult to determine. Therefore, no evidence exists to confirm the age of these dolomitized deposits as Quaternary Limestone Terraces rather than a part of the Seroe Domi Formation. Dolomite has also been reported from the Quaternary Terrace Limestones on Bonaire (Bandoian and Murray, 1974). However no detailed petrographic or geochemical analyses are reported by Bandoian and Murray (1974) with which to rigorously evaluate this interpretation or permit correlations with Dolomites I, I', and II.

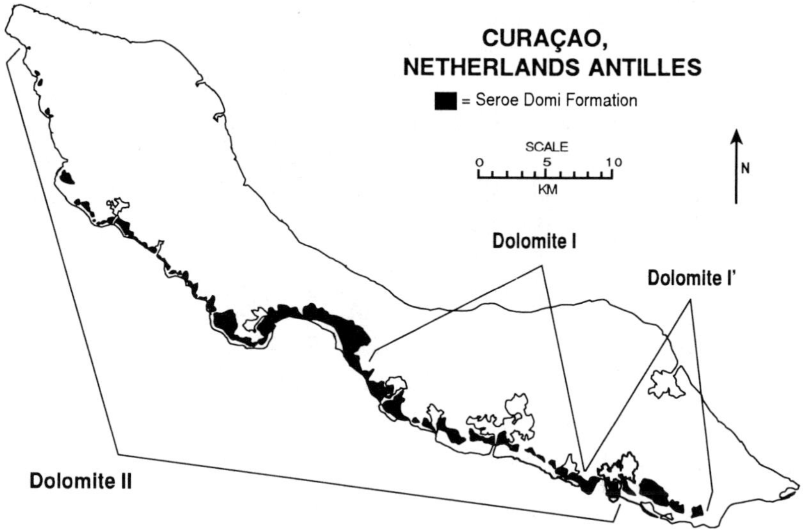

FIGURE 4.6. Geographic map of the island of Curaçao, Netherlands Antilles, exhibiting the distribution of Dolomites I, I', and II. The Seroe Domi Formation outcrop belt on the leeward coast of Curaçao is in black (after De Buisonjé, 1974).

Stratigraphic Distribution

A summary of the stratigraphic distribution of Dolomite I, I', and II within each outcrop of the Seroe Domi Formation on Curaçao is presented in Table 4.1 and Figs. 4.7 through 4.10. Detailed descriptions of the stratigraphic distribution of dolomite within representative Subunit 1 and 2 type sections are presented in this section in order to define the geometry of the dolomite bodies.

Dolomite I is primarily distributed as stratiform dolomite bodies (SDBs) within Subunit 1 limestones. The only exception to this is at Seru Sami, where the entire 24 m-thick deposit of Lithofacies 1 grainstones exhibits pervasive bed cross-cutting dolomitization. The stratiform dolomitization of Subunit 1 limestones is best exposed at Seru Blandan and St. Michielsberg (Localities 25 and 26 in Fig. 4.5; Fig. 4.11). Six SDBs have been observed within the lower 19 m of the Subunit 1 section at Seru Blandan, while five stratiform dolomite bodies occur at St. Michielsberg within the lower 18 m of the sec-

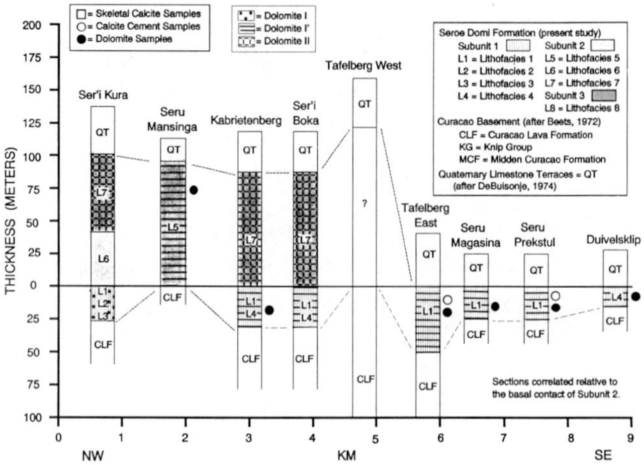

FIGURE 4.7. Distributions of Dolomite I, I', and II superimposed on a lithostratigraphic diagram of Curaçao from Ser'i Kura to Duivelsklip. Diagram illustrates the locality and lithofacies in which Dolomite I, I', and II were observed, and not the stratigraphic distribution within each lithofacies. Diagrams correlate the distribution of Subunits 1, 2, and 3 within the Seroe Domi Formation (present study), Cretaceous and Tertiary lithologies comprising the basement (after Beets, 1972), and the Quaternary Limestone Terraces (after De Buisonjé, 1974). The stratigraphic elevations of shell material, calcite cements and dolomites used for geochemical analyses are depicted with open squares, open circles, and darkened circles respectively.

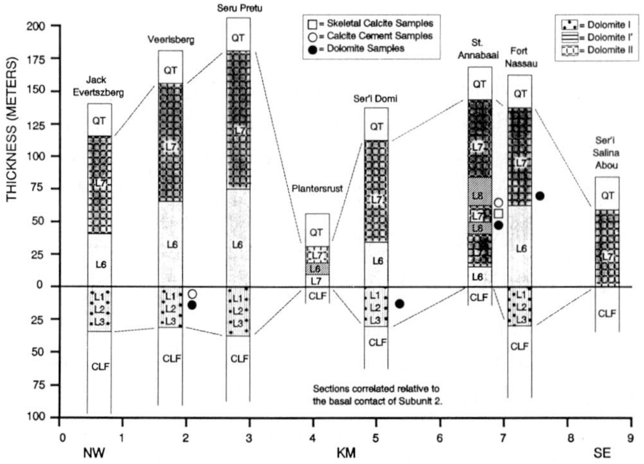

FIGURE 4.8. Distributions of Dolomite I, I', and II superimposed on a lithostratigraphic diagram of Curaçao from Jack Evertzberg to Ser'i Salina Abou. The description and legend are presented in Figure 4.7.

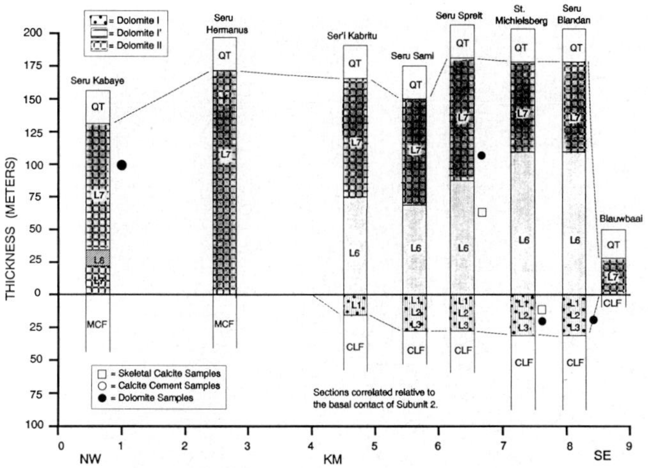

FIGURE 4.9. Distributions of Dolomite I, I', and II superimposed on a lithostratigraphic diagram of Curaçao from Seru Kabaye to Blauwbaai. The description and legend are presented in Figure 4.7.

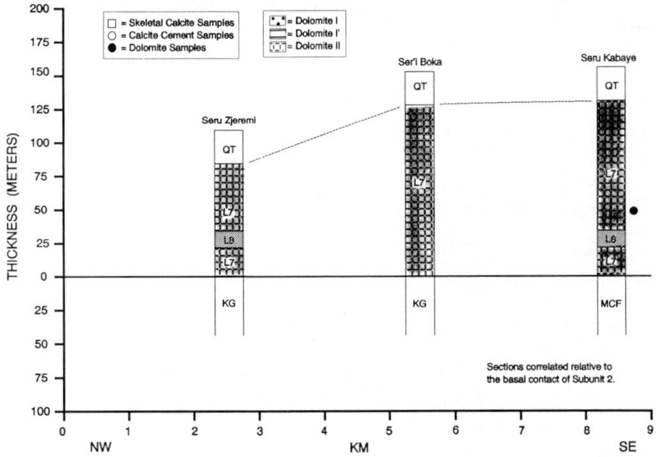

FIGURE 4.10. Distributions of Dolomite I, I', and II superimposed on a lithostratigraphic diagram of Curaçao from Seru Zjeremi to Seru Kabaye. The description and legend are presented in Figure 4.7.

FIGURE 4.11. Photographs of Dolomite I stratiform and bed cross-cutting pervasive bodies at St. Michielsberg. A. Stratiform dolomite bodies (SDB) 2 (2), 3 (3), and 4 (4). Note tan color of the dolomitized horizons in outcrop. B. Coalesence of SDB 4 (4) and 5 (5) westward of the allochthonous block approximately 20 m. Dashed line depicts upper and lower boundaries of dolomite bodies. C. Basal contact of Subunit 1 Lithofacies 1 grainstones with the Curaçao Lava Formation (CLF) at St. Michielsberg exhibiting fracture in the basement filled with carbonate muds and sands (shown with arrow). Photomicrographs of Dolomite I cements within the fractures shown in figure 3.20. D. SDB 6 (6) at Seru Blandan, showing upper and lower surfaces of SDB 6.

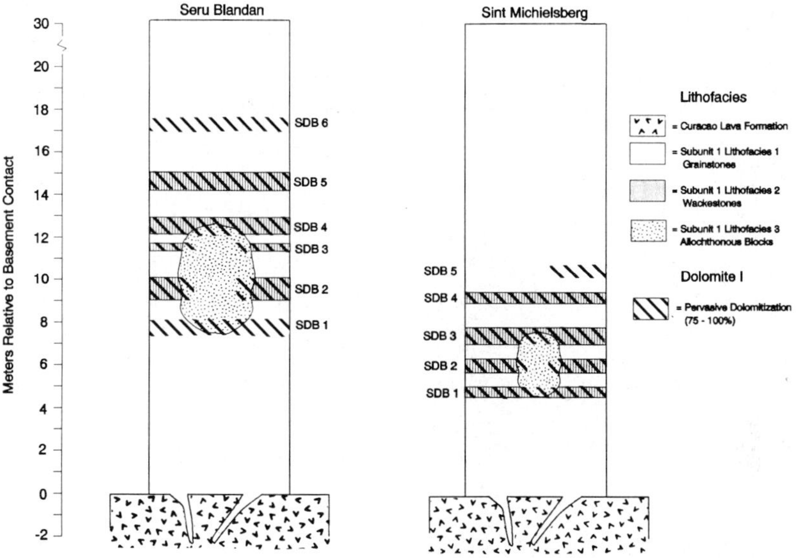

FIGURE 4.12. Fence diagram illustrating the elevation of Seru Blandan stratiform dolomite body (SDB) 1 through 6 with respect to St. Michielsberg SDB 1 through 5.

tion (Fig. 4.12). The limestones between SDBs vary from being partially to non-dolomitized, while limestones above the SDBs in the upper third of each Subunit 1 are non-dolomitized. However, the overall geometry of the SDBs remains a thin planar sheet approximately 1 m-thick that mimics the strike and dip of the host Subunit 1 limestone bedding planes. SDBs occur within Lithofacies 1 grainstones as well as Lithofacies 2 wackestones. Each SDB has been mapped along the length of each face of the flatiron outcrop to provide three-dimensional control on the distribution and shape of the dolomitized limestones (Figs. 4.13 and 4.14). These drawings illustrate that the SDBs are sometimes discontinuous down-dip from the allochthonous block that occurs at the apex of each flatiron, and either grade downdip into equivalent partially dolomitized lithofacies or are covered. Each SDB is surrounded by "halos" of pervasive to partially dolomitized limestones that extend into the base of overlying beds and top of underlying beds over distances of a few centimeters (Fig. 4.15). Dolomitization within Lithofacies 3 grainstones at both Seru Blandan and St. Michielsberg is confined to the outer 0.5 m rind of the allochthonous blocks (Figs. 4.13 and 4.14). Dolomite has also been observed

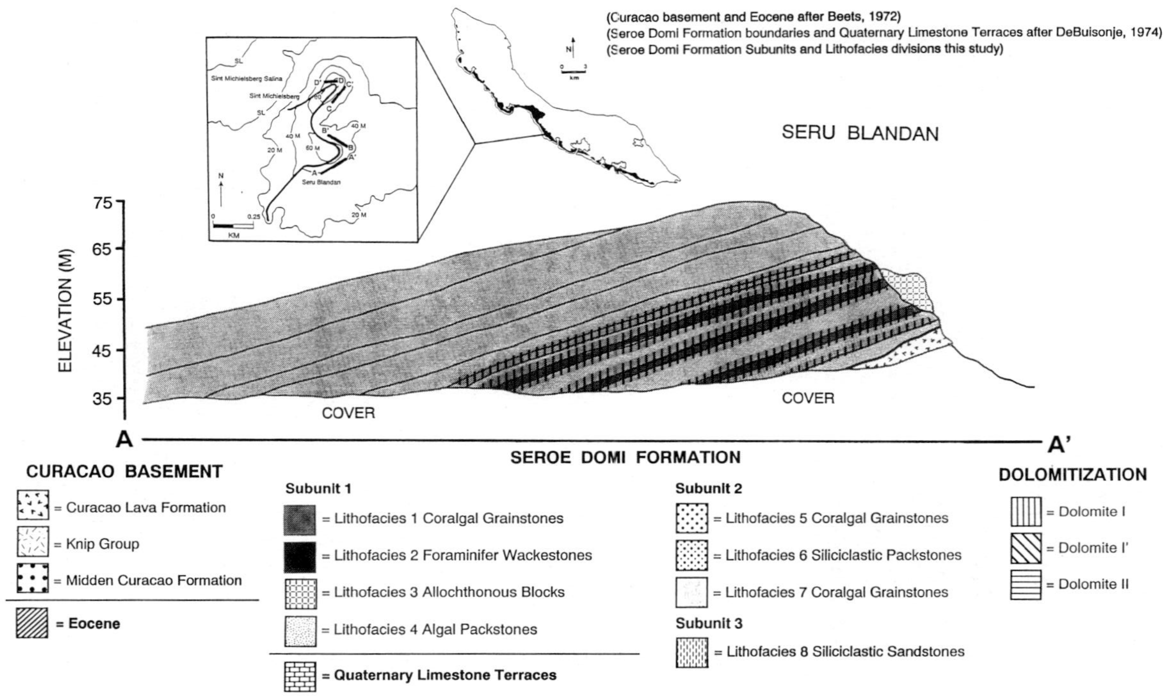

FIGURE 4.13. Distributions of Dolomite I (closely spaced vertical lines) superimposed on a line drawing of the southeastern face of Seru Blandan. Bedding and outcrop geometries traced from photograph in Figure 2.4A. Startiform dolomite bodies (SDB) 3 and 4 appear as one horizon at this scale.

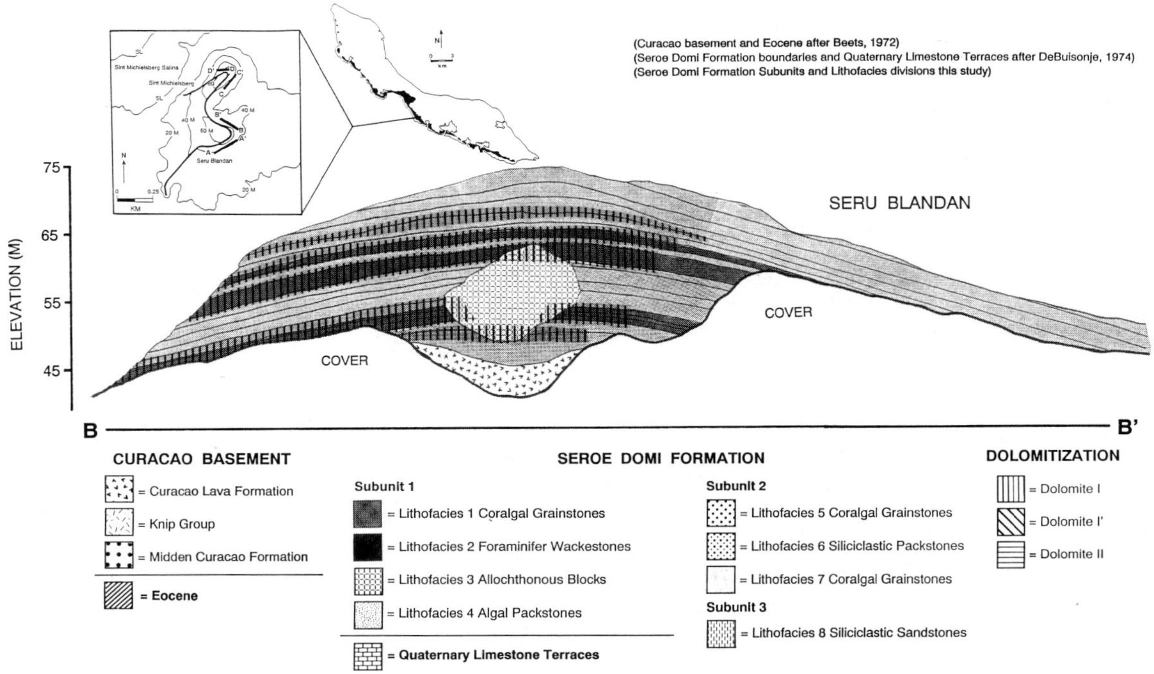

FIGURE 4.14. Distributions of Dolomite I (closely spaced vertical lines) superimposed on a line drawing of the northwestern face of Seru Blandan. Bedding and outcrop geometries traced from photograph in Figure 2.4B. Startiform dolomite bodies (SDB) 3 and 4 appear as one horizon at this scale.

in carbonate sediments that fill fractures within the Curaçao Lava Formation, at depths of as much as 2 m below the limestone/basement contact (Fig. 4.11).

Dolomite I' predominantly occurs as bed cross-cutting dolomite bodies that have pervasively dolomitized the entire outcrop. The Dolomite I' bed cross-cutting dolomitization of Subunit 1 is best exposed at Duivelsklip (Locality 33 in Fig. 4.5), where the entire 30-m thickness of Lithofacies 4 algal packstones and grainstones has been pervasively dolomitized. Dolomite I' also exhibits stratiform dolomitization in the Subunit 1 Lithofacies 1 deposits at Seru Prekstul (Locality 43 in Fig. 4.5), occurring as 0.25 to 1 m thick sheets of dolomite similar to those at Seru Blandan and St. Michielsberg. However, nowhere was Dolomite I' observed to have a local patchy distribution. The Seroe Domi Formation outcrop at Seru Mansinga (Locality 38 in Fig. 4.5) is composed of Subunit 2 Lithofacies 5, in which the entire deposit exhibits pervasive bed cross-cutting dolomitization by Dolomite I'.

Unlike Dolomites I and I', *Dolomite II* occurs as patchy, poorly developed localized areas of pervasive dolomitization that are distributed exclusively within Subunit 2 Lithofacies 7 grainstones. The type section for Dolomite II is at St. Annabaai, where patches of pervasive dolomitization are apparently randomly distributed with the middle 30 m of the section (Fouke, 1993). Similar patches of Dolomite II have also been observed at Seru Spreit, Seru Kabritu, and Seru Zjeremi (Localities 3, 14 and 24 in Fig. 4.5).

Correlation of Dolomite I, I', and II Cathodoluminesence

Crystal size and the CL-defined concentric crystal zonations observed within the Seroe Domi Formation dolomites are consistent for Dolomite I, I', and II throughout their respective geographic and stratigraphic distributions. Dolomite I samples from extreme updip, intermediate, and extreme downdip positions within SDB 4 at Seru Blandan show that the dolomite crystals are identical in size and CL, and exhibit no overgrowth trends. The same is true of comparisons between SDB 1, 2, 3, 4, 5, and 6 at Seru Blandan sampled from extreme updip, intermediate, and extreme downdip positions, with crystals being identical throughout the flatiron outcrop.

Dolomite-Limestone Diagenetic Contacts

The stratiform and bed cross-cutting dolomite bodies observed within the Seroe Domi Formation limestones exhibit distinct dolomite-limestone transition zones at their margins. These diagenetic contacts are composed of a gradation from pervasively dolomitized limestone within the dolomite bodies, to partially and non-dolomitized limestones comprising the surrounding host rock. Petrographic analysis of samples from vertically and horizontally oriented transects through these dolomite-limestone contacts provide "snap-shots" of the incipient stages of the dolomitization process, and provide a means to track the mineralogical and geochemical evolution of the fluid-rock system. Petrographic evidence of the timing and mode of dolomitization gained from these contacts will serve as a spatial and temporal frame of reference with which to evaluate the geochemistry of Dolomites I, I' and II.

The downdip dolomite-limestone transition zones of SDB 1 and SDB 2 were chosen as representative examples for analysis. SDB 1 is hosted by Lithofacies 1 grainstones that occur 8 m stratigraphically above the basement contact, and extends approximately 25 m downdip westward from the Lithofacies 3 allochthonous block before grading laterally into partially and nondolomitized grainstones (Fig. 4.15). Conversely, SDB 2 is hosted by Lithofacies 2 wackestones that lie 1 m above SDB 1, which extend approximately 18 m down-dip to the west and 125 m downdip to the east of the Lithofacies 3 allochthonous block before grading into partially and non-dolomitized limestones (Fig. 4.15). Therefore, comparison of SDB 1 and 2 transects permits the evaluation of the influence of facies selectivity on the processes and resulting fabric of dolomitization, as well as documentation of the intermediate steps of dolomitization.

Four sample transects, two perpendicular to bedding and two parallel to bedding, were taken across the down-dip regions of SDB 1 and 2 at Seru Blandan (Fig. 4.15). Each of the transects is 4 m in length and consists of 6 samples collected from the pervasively dolomitized stratiform dolomite bodies as well as from surrounding partially to non-dolomitized host limestones. Transects 1 and 3 remain within the SDB host limestone bed, while Transects 2 and 4 cross horizontal bedding planes and vertical facies transitions (Fig. 4.15). All four transects yield a consistent dolomitization fabric, which is herein called centripetal replacement dolomitization.

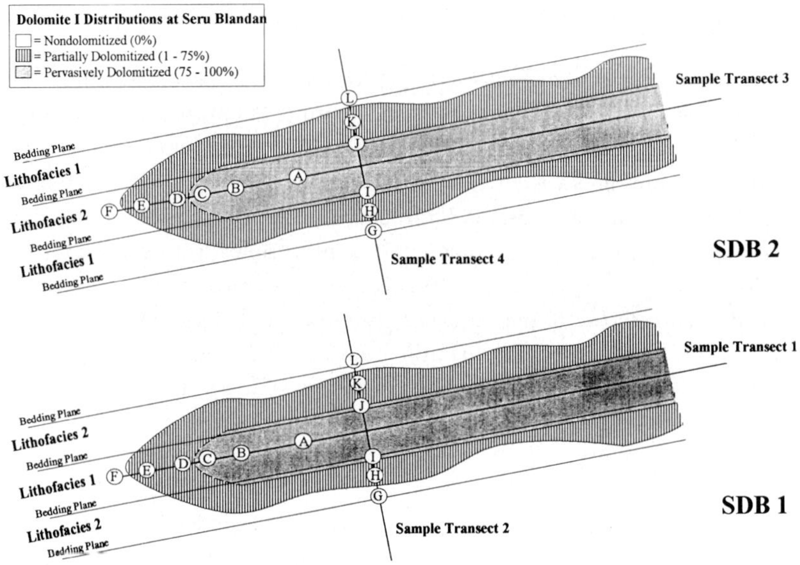

FIGURE 4.15. Schematic diagram of stratiform dolomite bodies (SDB) 1 and 2 at Seru Blandan. A and B. Diagrams depict geometry of the partially and pervasively dolomitized limestones with respect to bedding geometries, and the spatial orientation and position of sample transects 1, 2, 3, and 4.

The Centripetal Dolomitization Replacement Process

The correlation of concentric CL zonations in Dolomite I and I' crystals suggests that many individual dolomite rhombohedra are the product of both cement and replacement processes. The partially dolomitized host limestones in which these crystals occur permit the spatial distribution and geometric continuity of Dolomite I and I' CL zonations to be evaluated with respect to the original grains and pore space in which they were precipitated. This provides unequivocal evidence that cementation pre-dated replacement. Dolomite cementation and replacement occur in compositionally different types of diagenetic fluids, the first being saturated with respect to calcite and the second being undersaturated with respect to calcite. Therefore, these unique petrographic relationships provide evidence that there was a significant change in fluid composition during the dolomitization process in the Seroe Domi Formation. To illustrate these petrographic relationships,

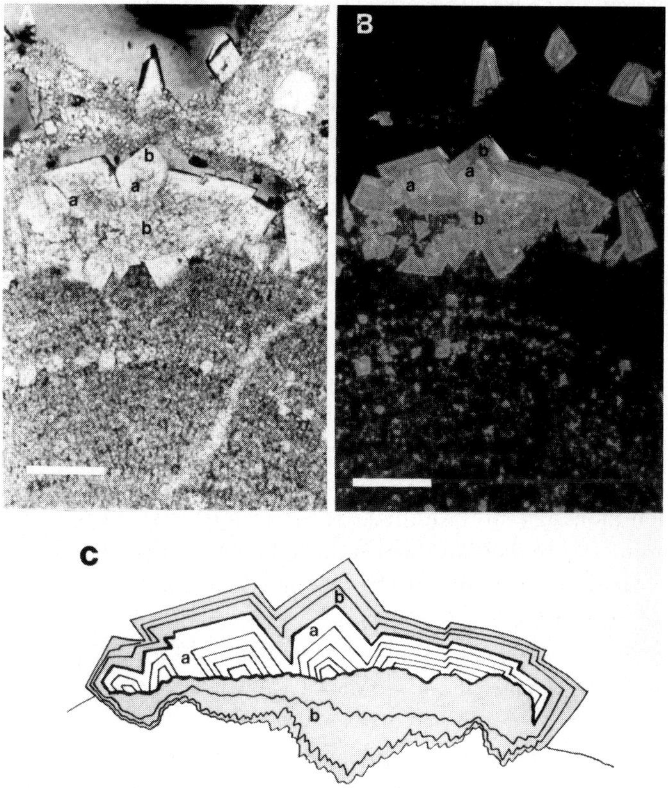

FIGURE 4.16. Petrography of the centripetal Dolomite I' replacement process in a red algal fragment. Petrographic analysis of sample number 2/25/89-III1 from Subunit 1 Lithofacies 1 grainstones at Seru Prekstul. A and B. Paired plane-light and CL photomicrographs of Dolomite 1 cements growing into, and centripetally replacing the red algal. An initial light orange CL dolomite cement phase (a) grows against the grain edge and exhibits truncated concentric CL crystal zones. This was followed by a second darker orange CL dolomitization phase (b), which included both cementation and replacement of the algal grain. Planar concentric CL crystal zones in cement crystals (the b toward the top of photograph) continue into the red algal fragment as irregular non-planar zones (the b toward the top of photograph). Scale bars are 75 μm. C. Traced line drawing of photograph B depicting the geometries of concentric CL banding. The initial cement dolomite phase (a) is white, while the second dolomite cement and replacement phase (b) is shaded. Prominent concentric CL zonations have been chosen from photograph B and placed on this sketch in order to illustrate the truncation of zones against the grain margin in phase a, and the continuity of zones that change from planar to non-planar geometries in the cement to replacement portions of each crystal respectively. Note that the dolomitization fronts (the margins of crystals) are planar for cements and a mixture of planar and irregular for replacement dolomite.

the partial dolomitization of red algae by Dolomite I' and benthic foraminifers by Dolomite I will be described.

In plane light, Dolomite I and I' crystals are composed of two components. These include a euhedral cement portion that encrusts the outer surface of grains and extends into surrounding pore-space, and a replacement portion that extends into the skeletal material as well as surrounding blocky and bladed calcite cements (Figs. 4.16 and 4.17). In CL, concentric zonations within the core of each Dolomite I' crystal extending in void space are observed to abut against and be truncated by the outer surface of the skeletal grains (Fig. 4.16). This indicates that these are cements that grew along the edge of the red algae grains *without* replacing the grains. Conversely, concentric CL zones within the outer portions of the Dolomite I' crystals that grow in void space are observed to continue into the replacement portion of each crystal (Fig. 4.16). Individual concentric zones are observed to continue undisrupted from being planar features in the dolomite crystals in pore space, to irregular non-planar features within the red algal grain (Fig. 4.16). In the benthic foraminifers, pre-dolomite bladed calcite cements are incorporated within the replacement dolomite as elongate inclusions or crystal molds (Fig. 4.17) around which CL zones continue undisrupted.

The petrographic relationships documented in Figs. 4.16 and 4.17 suggest that dolomitization took place in two chronologically distinct phases. In the first phase, dolomite cementation occurred without accompanying replacement dolomitization, suggesting the fluids were saturated with respect to calcite. In the second phase, the continuity of CL concentric zones in cement and replacement crystals suggests that dolomite cementation and replacement were synchronous. Therefore dolomite replacement in these circumstances is the result of a centripetal process in which grains were initially replaced at the outer margins, with ensuing replacement progressing toward the center of the grains. In addition, the resultant replacement CL zones are identical, regardless of whether it was formed by replacement or cementation. Replacement dolomitization has also been observed to occur randomly within grains, and is not always part of a spatial progression from the rim to the center of grains (Fig. 4.16). However, the CL zones within these replacement dolomites are always consistent with the intensity, color, and hairline zoning of the CL zones in the outer rinds of the dolomite cements. Unfortunately, similar interpretations are not possible within the pervasively dolomitized stratiform dolomite bodies because the pre-dolomitization

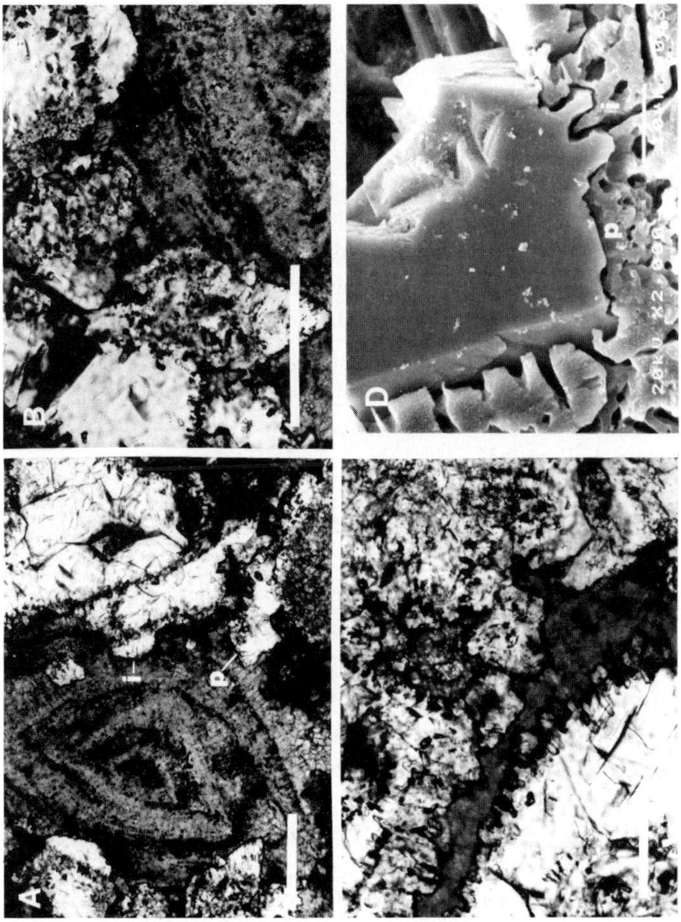

FIGURE 4.17. Petrography of the centripetal replacement of a benthic foraminifer by Dolomite I, illustrating the incorporation of pre-dolomite bladed calcite cements as inclusions. Petrographic analysis of sample number 2/25/89-III1 from Subunit 1 Lithofacies 1 grainstones at Seru Prekstul. Samples are stained with Alizarin-red. A, B, and C. Plane-light photomicrographs of Dolomite I centripetally replacing the margins of a benthic foraminifer. Unlike in the red algae (Figure 5.19), the replacement dolomitization front is both planar (p) and irregular (i). The pre-dolomite bladed calcite cements are incorporated as either: 1) linear to curvilinear inclusion arrays in their growth positions; or 2) random unaligned inclusions apparently disrupted from their growth positions. D. Scanning electron micrograph image of the contact between the benthic foraminifer and the replacement dolomite. Sample was etched in 1% formic acid for 10 seconds. Note that dolomitization front is both planar (p) and irregular (i). Scale bar in A is 200 μm. Scale bar in B is 200 μm. Scale bar in C is 125 μm.

limestone fabric is no longer present. Therefore no frame of reference exists with which to evaluate the planar versus non-planar orientations of the Dolomite I and I' Cl zonations.

Seroe Domi Formation Host Limestone Diagenesis

The Seroe Domi Formation limestones containing Dolomites I, I', and II have experienced several diagenetic episodes of leaching, calcitization, and dolomitization, which took place after each successive Subunit 1, 2, and 3 depositional period. This has created a complex paragenetic sequence of depositional and diagenetic events that are summarized in Fig. 4.18. This sequence provides a frame of reference with which to interpret the precipitation of Dolomites I, I', and II.

Pre-Dolomite I and I' Diagenesis

All Subunit 1 Lithofacies 1, 2, 3, and 4 limestones experienced extensive leaching of skeletal grains and calcite cementation prior to dolomitization. Aragonitic corals, bivalves, and gastropods are intensely leached and often completely dissolved, leaving biomolds and micrite envelopes within which dolomite cements have been precipitated (Fig. 4.19). However, calcitic skeletons of red algae, foraminifers, and echinoderms are generally better preserved but also exhibit highly variable extents of dissolution. These fabrics range from well-preserved to partially or completely dissolved specimens within a single thin-section.

A distinct pre-Dolomite I and I' bladed calcite cement 25 to 50 μm in length (after terminology of Folk, 1965) post-dates limestone dissolution within Subunit 1 Lithofacies 1, 2, and 3 (Fig. 4.20). These bladed calcite cements vary from patchy isolated crystals to locally continuous cement rims that are best developed on the tests of low-Mg calcite pelagic foraminifers. In addition, these bladed calcite cements exhibit bright hairlined CL zonations, with larger crystals displaying an additional outer zone of non-luminescent cement (Fig. 4.20). Dolomite I and I' often encrust and encase these bladed calcite cements and generally preserve their original morphology and CL fabrics (Fig. 4.20). However, some of the dolomite-encased bladed calcites have developed jagged irregular surfaces and a mottled CL, while others are

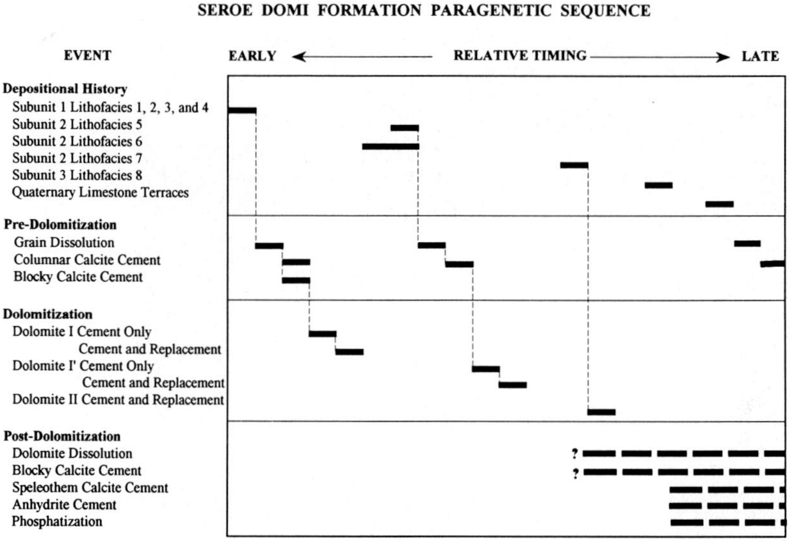

FIGURE 4.18. Seroe Domi Formation paragenetic sequence. The relative timing of Dolomite I and I' dissolution and post-Dolomite I and I' calcite cementation is uncertain.

completely dissolved leaving crystal molds. These petrographic relationships between the bladed calcites and Dolomite I and I' are identical to predolomite bladed cements described in the Seroe Domi Formation on Bonaire by Sibley (1980).

Blocky non-luminescent calcite cements 5 to 75 μm in diameter also predate Dolomite I in Subunit 1 lithologies, as evidenced by their replacement with Dolomite I in the outer margins of Lithofacies 3 allochthonous blocks. These blocky calcites are most abundant and best developed within Lithofacies 3 allochthonous grainstone blocks, where they encrust the early bladed calcite cements and occlude the remaining intergranular pores. Dolomitization is restricted to the outer 0.25 m rims of these tight allochthonous blocks, and occurs either within intraskeletal pores or as local replacements of skeletal grains, cement-grain contacts, and bladed and blocky calcite cements. The Lithofacies 3 bladed and blocky calcite cements are similar in crystal morphology and CL to those observed in Lithofacies 1, 2, and 4, which suggests that they are the same generation of cement (Fouke, 1993).

Subunit 1 Lithofacies 4 and Subunit 2 Lithofacies 5 have experienced pervasive fabric-destructive dolomitization, and therefore no partially dolomi-

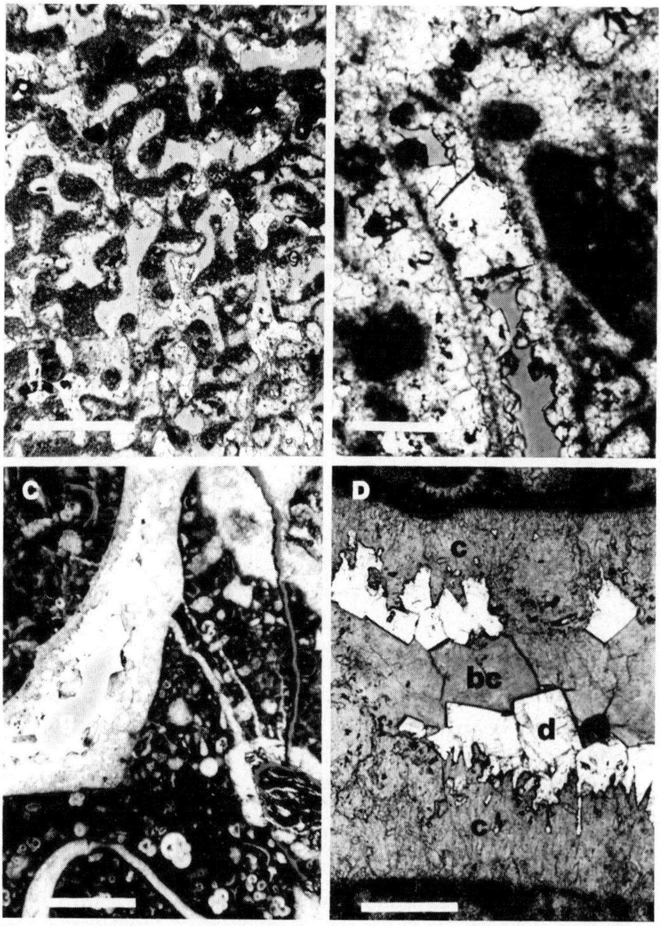

FIGURE 4.19. Petrographic evidence of extensive leaching of skeletal grains and calcite cementation prior to Dolomite I and I' precipitation. A. Plane-light photomicrograph of Dolomite I crystals precipitated within coral molds in Subunit 1 Lithofacies 1 grainstones at Seru Blandan. Stained with Alizarin-red. Sample number 3/30/89-III2. Scale bar is 100 μm. B. Plane-light photomicrograph of Dolomite I crystal precipitated within a bivalve mold in Subunit 1 Lithofacies 1 grainstones from St. Michielsberg. Stained with Alizarin-red. Sample number 2/15/89-III3. Scale bar is 0.5 mm. C and D. Plane-light photomicrographs of a gastropod mold within carbonate sediments filling fractures within the Curaçao Lava Formation approximately 2 meters below the Subunit 1 Lithofacies 1 contact at St. Michielsberg. Photograph D is an enlargement of the mold near the top of the picture in C, which has been rotated 90° to the left. Dolomite I crystals (d) encrust bladed calcite cements (bc) and are in turn encrusted by blocky calcite cements (c). Stained with Alizarin-red. Sample number 2/15/89-III1. Scale bar is 400 μm in C and 100 μm in D.

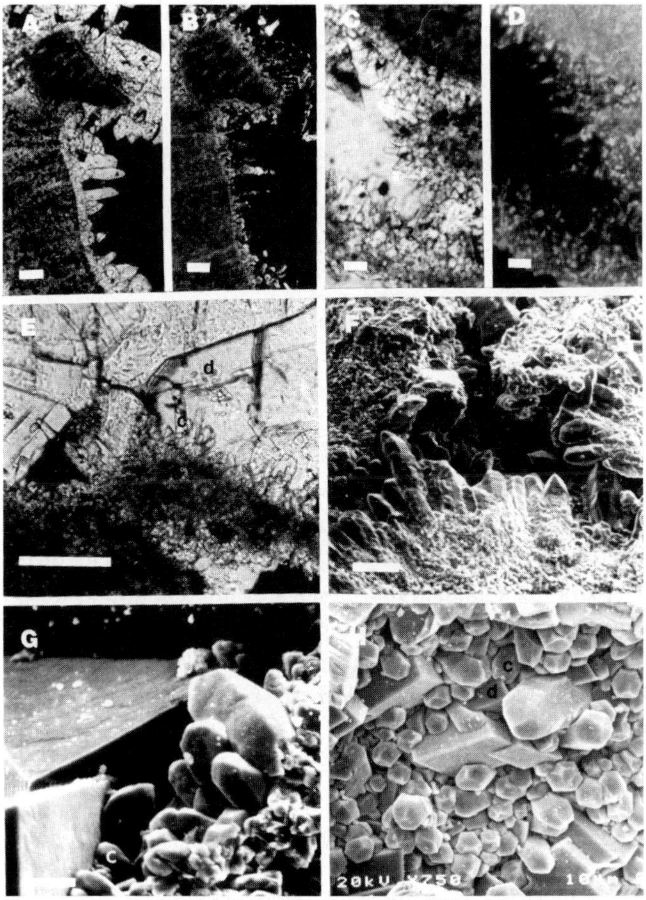

FIGURE 4.20. Petrography of pre-Dolomite I and I' bladed calcite cements. A and B. Paired plane-light and CL photomicrographs of pre-Dolomite I bladed calcite cements from Subunit 1 Lithofacies 1 grainstones at St. Michielsberg. Sample number 2/14/89-II1. Scale bars are 10 μm. C and D. Paired plane-light and CL photomicrographs of bladed calcite cements from a Subunit 1 Lithofacies 3 allochthonous block at St. Michielsberg. Similarity in crystal morphology, size, and CL character suggest that the bladed calcites in Lithofacies 3 are equivalent to those in Lithofacies 1. Sample number 2/15/89-II13. Scale bars are 10 μm. E. Plane-light photomicrograph of pre-dolomite bladed calcite cements (c) encrusted by Dolomite I rhombohedra (d). From sample number 2/14/89-II1 in Subunit 1 Lithofacies 1 grainstones at St. Michielsberg. Scale bar is 50 μm. F. Scanning electron micrograph of pre-dolomite bladed calcite cements from the sample shown in photograph E. Scale bar is 10 μm. G and H. Scanning electron micrograph of pre-dolomite bladed calcite cements (c) encrusted by Dolomite I rhombohedra (d). From sample number 2/25/89-III3 in Subunit 1 Lithofacies 1 grainstones at Seru Prekstul. Scale bar in G is 10 μm.

tized limestones were available to reconstruct the pre-dolomitization history as was done in Lithofacies 1, 2, and 3. However, bladed calcite cements have been preserved as inclusions within Dolomite I’ crystals during centripetal dolomitization, creating linear to curvilinear inclusion trains that preserve the margins of precursor grains. These bladed calcite inclusions are identical to pre-Dolomite I and I’ bladed calcite cements, and suggest that Lithofacies 4 and 5 also experienced an early diagenetic history of leaching and calcitization similar to that observed in Lithofacies 1 and 2.

Pre-Dolomite II Diagenesis

No pre-Dolomite II bladed calcite cements occur within Subunit 2 Lithofacies 7 grainstones as they do in Subunit 1 and Subunit 2 Lithofacies 5. In addition, no petrographic relationships were observed to determine the relative timing of leaching and skeletal alteration with Dolomite II precipitation. Unlike Dolomite I’ crystals in Lithofacies 4 and 5, Dolomite II crystals contain small (< 5 m), irregularly shaped, randomly distributed calcite inclusions that do not exhibit the bladed morphology of pre-dolomite cements (Fig. 4.4). In addition, non-dolomitized samples of Lithofacies 7 contain no bladed or blocky calcites similar in either morphology or CL to pre-Dolomite I and I’ bladed and blocky calcites. Therefore, the relative timing of Lithofacies 7 grainstone stabilization with respect to Dolomite II precipitation is uncertain.

Post-Dolomite I, I’, and II Diagenesis

Several post-Dolomite I, I’, and II diagenetic fabrics and phases have been observed within the Seroe Domi Formation (Fig. 4.18). Because the Dolomite I, I’, and II samples chosen for analysis exhibit no evidence of recrystallization, post-dolomitization diagenesis is interpreted not to have significantly affected the geochemical composition of Dolomite I, I’, and II.

A small percentage of Dolomite I and I’ crystals exhibit selective leaching of their cores and concentric outer crystal zones. These leached crystals are also sometimes filled with a non-luminescent blocky calcite cement. Based on coordinated plane-light and SEM petrography, dissolution appears to have proceeded along irregular conduits that cut perpendicular to the crystal faces to form a porous network through the entire thickness of each crystal. How-

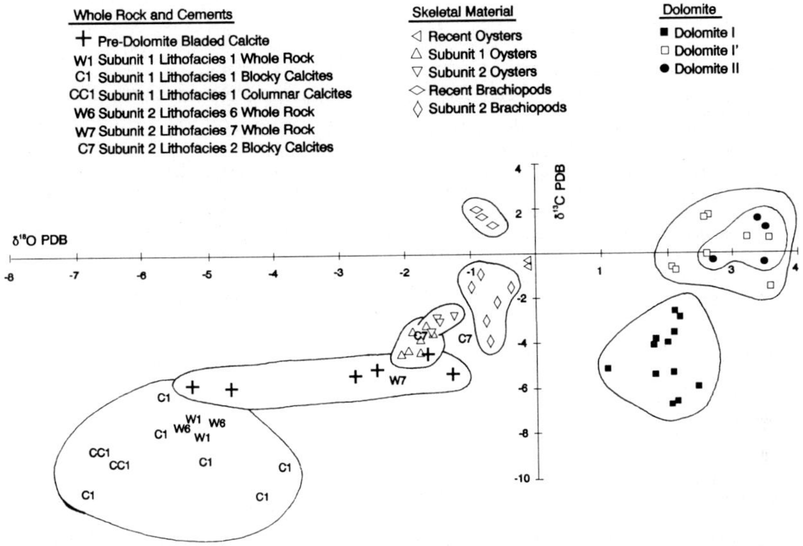

FIGURE 4.21. Composite cross-plot of $\delta^{18}O$ versus $\delta^{13}C$ composition of skeletal material, calcite cements, whole-rock analyses, and dolomite from the Seroe Domi Formation.

ever the remaining portion of each partially dissolved crystal retain their original concentric CL zoning (Fouke, 1993). Dolomite I and I' dissolution occurs in random isolated patches, and exhibits no systematic lithologic or stratigraphic distributions within the formation.

GEOCHEMISTRY OF SEROE DOMI FORMATION DOLOMITES AND CALCITES

The geochemical analyses of Seroe Domi Formation skeletal material, calcite cements, and Dolomites I, I', and II are presented in this section.

Skeletal Material and Pre- and Post-Dolomitization Calcites

Geochemical analyses of Subunit 1 and 2 oyster and brachiopod shells and pre- and post-dolomitization calcites are presented in Tables 3.1 and 4.2. These analyses suggest that all pre- and post-dolomite calcite cements are composed of low-Mg calcite (< 3 mole % $MgCO_3$) and are depleted in $\delta^{13}C$ and $\delta^{18}O$ (Fig. 4.21). Laser ablation analyses (LASSIE) of the Subunit 1 pre-

Sampling Locality	Sample Number	Cement Description	Mg ppm	Mg mol%	Sr ppm	1/Sr ppm	Sr/Ca 1000x	Mn ppm	Fe ppm	δ¹⁸O PDB	δ¹³C PDB
		Standard Deviation	85	0.04	6	0.17		7	51	0.10	0.20
		Detection Limit	69	0.03	2	0.50		22	54		
Subunit 1 Lithofacies 1											
St Michielsberg	2/14/89-II2	Bladed Calcite	1938	0.8	109	0.009	0.27	285	227	-2.41	-5.16
St Michielsberg	2/14/89-II2	Bladed Calcite	1927	0.8	62	0.016	0.16	167	138	-2.75	-5.41
St Michielsberg	2/14/89-II2	Bladed Calcite	1948	0.8	19	0.053	0.05	123	102	-1.26	-5.37
St Michielsberg	2/14/89-II2	Bladed Calcite	1954	0.8	70	0.014	0.18	189	159	-5.24	-5.80
St Michielsberg	2/14/89-II2	Bladed Calcite	1234	0.5	23	0.043	0.06	35	120	-4.65	-5.97
St Michielsberg	2/14/89-II2	Bladed Calcite	2062	0.8	61	0.016	0.15	207	115	-1.65	-4.47
Average Composition			1844	0.8	57	0.025	0.14	168	144	-2.14	-5.31
St Michielsberg	2/15/89-V7	Blocky Calcite	3119	1.3	76	0.013	0.19	305	122	-6.55	-8.77
St Michielsberg	2/15/89-V7	Blocky Calcite	1101	0.5	9	0.111	0.02	bd	bd	-3.84	-9.39
St Michielsberg	2/15/89-V7	Blocky Calcite	3221	1.3	33	0.030	0.08	205	bd	-5.06	-9.09
St Michielsberg	2/15/89-V7	Blocky Calcite	976	0.4	116	0.009	0.29	183	221	-5.66	-6.26
St Michielsberg	2/15/89-V7	Whole Rock								-5.13	-7.65
St Michielsberg	2/15/89-V7	Whole Rock								-5.47	-7.01
Seru Prekstul	2/27/89-IV7	Blocky Calcite								-4.20	-5.63
Seru Prekstul	2/27/89-IV7	Blocky Calcite								-5.72	-7.88
Tafelberg	7/1/88-I	Blocky Calcite								-6.32	-9.24
Tafelberg	7/1/88-I	Blocky Calcite								-6.85	-10.58
Tafelberg	7/1/88-I	Blocky Calcite								-6.33	-9.24
Average Composition			2104	0.9	59	0.041	0.15	231	172	-5.56	-8.25
Subunit 2 Lithofacies 6											
Seru Spreit	9/16/90-IIb	Whole Rock								-5.86	-7.05
Seru Spreit	9/16/90-IIb	Whole Rock								-4.91	-7.45
Average Composition										-5.39	-7.25
Subunit 2 Lithofacies 7											
St Annabaai	9/15/90-1A	Blocky Calcite	7718	3.2	145	0.007	0.37	365	195	-1.07	-3.81
St Annabaai	9/15/90-1A	Blocky Calcite	2396	1.0	56	0.018	0.14	bd	bd	-1.77	-3.66
St Annabaai	9/15/90-1A	Whole Rock								-2.09	-5.62
Average Composition			5057	2.1	101	0.012	0.26	365	195	-1.64	-4.36

Dolomite I and II compositions presented in Chapter 3

bd = below detection
Ca, Mg, and Fe determined by electron probe microanalysis (EPMA).
Sr and Mn determined by direct coupled plasma (DCP) analysis.

TABLE 4.2. The geochemical composition of Seroe Domi Formation calcite cements and whole rock analyses.

dolomite bladed calcite cements suggest that they are composed of a broad range of depleted $\delta^{18}O$ values (−1.26 to −5.24‰ PDB) but a narrow range of depleted $\delta^{13}C$ compositions (−4.47 to −5.97‰ PDB; Fig. 4.21). Micro-sampled post-dolomite blocky and columnar calcite cements (designated as Subunit 1 and 2 blocky and columnar calcites in Fig. 4.21) are significantly more depleted in $\delta^{18}O$ and $\delta^{13}C$ than the pre-dolomite bladed calcites, ranging from −3.84 to −6.85‰ PDB $\delta^{18}O$ and −5.31 to −10.58‰ PDB $\delta^{13}C$ (Fig. 4.21).

Sample Number	Locality		Ca Mole%	Mn ppm	Na ppm	Sr ppm	Cl ppm	F ppm	S04 ppm	$\delta^{18}O$ PDB	$\delta^{13}C$ PDB	$^{87}Sr/^{86}Sr$	Zn ppm	Cu ppm
		Stand. Dev.	1.3	7	5	6	15	15	100	0.10	0.20	0.000020	2	2
		Detect. Limit		22	7	2	80	90	300				3	3
	Dolomite I													
1	Seru Blandan	2/21/89-I3N	55.7	354	234	251	*bd*	350	786	2.08	-5.40	0.708875		
2	Seru Blandan	2/13/89-I3N	55.6	942	238	112	207	323	794	1.04	-5.18	0.708875		
3	Seru Blandan	3/29/89-II5	55.6	264	239	236	210	320	800	2.00	-4.02	0.708891		
4	Seru Blandan	3/30/89-IV3	54.3	306	236	245	*bd*	321	899	2.06	-3.51	0.708880		
5	Seru Blandan	3/29/89-II7	55.9	422	246	231	243	352	788	2.15	-2.84	0.708885		
6	Seru Blandan	3/30/89-V2	55.1	386	233	242	187	300	804	1.80	-3.94	0.708895		
7	Seru Blandan	2/21/89-II5	55.7	369	250	202	233	275	901	1.78	-5.42	0.708886		
8	St. Michielsberg	2/14/89-II5	54.5	271	196	188	370	307	734	2.05	-6.79	0.708866		
9	St. Michielsberg	2/14/89-II4	54.6	295	210	198	191	279	821	2.14	-6.62	0.708829		
10	St. Michielsberg	2/14/89-II9	54.9	358	206	175	*bd*	286	804	2.46	-6.00	0.708859	*bd*	*bd*
11	Veerisberg	5/23/88-II7	55.0	117	241	232	243	270	944	2.11	-2.55	0.708880		
12	Seru Domi	5/27/88-III1	56.1	649	232	252	*bd*	343	853	1.76	-4.15	0.708879		
	Average Composition		55.3	394	230	214	236	311	827	1.95	-4.70	0.708875		
	Dolomite I'													
13	Duivelsklip	2/26/89-I3	54.5	138	276	160	*bd*	452	912	2.64	1.69	0.708989		
14	Seru Prekstul	2/25/89-III1	55.6	1434	332	231	96	461	714	2.11	-0.75	0.708991	*bd*	*bd*
15	Seru Prekstul	3/31/89-IX2	55.9	654	349	209	*bd*	500	672	3.22	0.68			
16	Seru Magasina	3/31/89-IX1	53.6	462	316	182	*bd*	592	948	2.60	-0.02			
17	Seru Mansinga	6/25/88-V3	56.1	110	251	186	98	390	874	3.58	0.71	0.708961		
18	Tafelberg	7/1/88-V7	54.4	300	272	162	*bd*	451	1107	2.56	1.53	0.709010		
19	Kabrietenberg	6/21/88-V3	54.8	272	219	189	379	267	680	2.08	-0.94			
20	Kabrietenberg	6/21/88-III8	54.7	1487	319	136	576	341	696	3.55	-1.53			
	Average Composition		55.0	607	292	182	287	432	825	2.79	0.17	0.708988		
	Dolomite II													
21	Seru Spreit	9/16/90-IIA	53.5	112	248	109	119	254	731	3.51	1.13	0.708934	20	20
22	St. Annabaai	4/5/89-I	51.2	95	223	182	*bd*	441	*bd*	2.69	-0.33	0.708968		
23	Seru Kabritu	5/22/90-II7	55.3	411	278	183	*bd*	497	1159	3.38	1.53	0.708990		
24	Seru Kabaye	5/11/90-IA	53.2	293	275	164	265	374	829	3.24	0.62	0.709052		
25	Fort Nassau	5/30/88-I7	55.7	554	283	152	470	377	626	3.49	-0.34	0.709068	*bd*	*bd*
	Average Composition		53.8	293	261	158	285	389	836	3.26	0.52	0.709002		

Zn and Cu abundances determined by synchrotron x-ray flouresence (SXRFA). *bd = below detection; na = not analyzed*
Cl, F and SO4 abundances determined by ion chromatography (IC).
Ca and Mg abundances determined by electron probe microanalyses (EPMA).
Na, Sr, Mn and Fe abundances determined by direct coupled plasma (DCP).

TABLE 4.3. The geochemical composition of Dolomites I, I', and II from the Seroe Domi Formation on Curaçao. A constant of 0.00002 was subtracted from the Sr isotope data in this study for normalization with the analyses of Hodell *et al.* (1991).

Geochemistry of Dolomites I, I', and II

Major and Trace Elements: Twelve samples of Dolomite I, eight samples of Dolomite I', and five samples of Dolomite II were geochemically analyzed to determine their major and trace element compositions (Table 4.3). These analyses suggest that Dolomite I, I', and II are relatively equivalent in their Ca, Sr, Na, Mn, Fe, F, Cl, SO_4, and $\delta^{18}O$ compositions, but distinctly different with respect to their Zn, Cu, $\delta^{13}C$ and $^{87}Sr/^{86}Sr$ compositions (Table 4.4). All three dolomites are Ca-rich , with Dolomite I containing 54.3 to 56.1 mole % $CaCO_3$, Dolomite I' containing 54.4 to 56.1 mole % $CaCO_3$, and Dolomite II containing 51.2 to 55.7 mole % $CaCO_3$. In addition, Dolomites I, I', and II exhibit similar low abundance compositional ranges of Sr (109-252 ppm),

	Estimated Seawater Dolomite	Dolomite I		Dolomite I'		Dolomite II	
		Mean	Range	Mean	Range	Mean	Range
Sr ppm	75 - 184	214	112 - 252	182	160 - 231	158	109 - 183
Mn ppm	19 - 22	394	271 - 942	607	110 - 1487	293	95 - 554
Fe ppm	16 - 57	750	240 - 1673	1445	185 - 4177	1924	536 - 4151
Na ppm	150 - 375	230	196 - 241	292	219 - 332	261	223 - 283
Zn ppm	< 2	< 2	< 2	< 2	< 2	10	< 2 - 20
Cu ppm	< 2	< 2	< 2	< 2	< 2	10	< 2 - 20
Cl ppm	157 - 494	180	<80 -370	166	<80 - 576	285	<80 - 470
F ppm	80 - 1500	311	275 - 343	432	267 - 592	389	254 - 497
SO4 ppm	232 - 2625	827	734 - 944	825	672 - 1107	836	< 300 - 1159
$\delta^{18}O$ PDB	+3.0 - +5.8	1.95	+1.04 - +2.46	2.79	+2.08 -+3.55	3.26	+2.69 - +3.51
$\delta^{13}C$ PDB	+0.5 - +2.5	-4.7	-2.55 - -6.79	-0.58	-1.53 - +1.69	0.52	-0.34 - +1.53
$^{87}Sr/^{86}Sr$	0.7087 - 0.7092	0.708875	0.708866 - 8915	0.708988	0.708981 - 9030	0.709002	0.708954 - 9088

TABLE 4.4. Average and range of geochemical compositions of Dolomites I, I', and II from the Seroe Domi Formation and the estimated seawater dolomite composition. A constant of 0.00002 was subtracted from the Sr isotope data in this study for normalization with the analyses of Hodell *et al.* (1991).

Rb (<1ppm), Na (196-332 ppm), Cl (<576 ppm), F (254-592 ppm), and SO_4 (<1159 ppm). Conversely, Mn exhibits large and erratic variations, which range from 271 to 942 ppm Mn in Dolomite I, 110 to 1487 ppm Mn in Dolomite I', and 95 to 554 ppm Mn in Dolomite II. Reconnaissance analyses of Dolomite I, I', and II with SXRFMA were also completed to analyze for Zn and Cu abundances. SXRFMA spectra were run for 10 crystals from a single thin-section for four different samples. The results suggest that Dolomite I, I', and a Dolomite II sample with the most radiogenic $^{87}Sr/^{86}Sr$ composition have Zn and Cu abundances that are below detection (<2 ppm; Fig. 4.22). In distinct contrast, a Dolomite II sample with the most nonradiogenic $^{87}Sr/^{86}Sr$ had compositions of 20 ppm Zn and 20 ppm Cu (Table 4.4; Fig. 4.22).

EPMA spot analyses along transects parallel and perpendicular to the distinct concentric CL crystal zonations observed in Dolomite I, I', and II are presented in Fouke and Reeder (1992). The results show that the $CaCO_3$ content is consistent along transects perpendicular to the growth direction within a single crystal, while some variation exists between crystals. However, differences of up to 3 mol% Ca have been observed between coeval growth sectors (Fouke and Reeder, 1992). Fe and Mn were analyzed at the same locations as were Ca and Mg in each Dolomite I, I', and II crystal. In transects along the growth direction within any given sector, Mn and Fe were found to be highly variable. Fe in particular shows large variations, ranging from below detection (approx. 65 ppm) to over 3000 ppm. Variations for Mn were less extreme, although still significant.

Oxygen, Carbon, and Strontium Isotopes: Dolomite I exhibits a compositional range in $\delta^{18}O$ of +1.04 to +2.46‰ PDB (mean = 1.95‰), while Dolomite I' and II have a similar overlapping range in $\delta^{18}O$ values of 2.08 to 3.55‰ PDB (Dolomite I' mean = 2.79‰; Dolomite II mean = 3.27‰; Table 4.4; Fig. 4.21). Conversely, Dolomite I exhibits a broad depleted range in $\delta^{13}C$ of −2.55 to −6.79‰ PDB (mean = −4.7‰), while Dolomites I' and II have tightly clustered enriched compositions of −1.53 to 1.69‰ PDB (Dolomite I' mean = −0.58; Dolomite II mean = 0.50; Table 4.4; Fig. 4.21). This plot illustrates that Dolomite I depletions in $\delta^{13}C$ are approximately equivalent to the depletions observed in the pre-Dolomite I bladed calcite cements and the Subunit 1 oysters (Fig. 4.21). In addition, the pre-dolomite bladed calcites and Subunit 1 and 2 unaltered and altered skeletal material are consistently depleted in $\delta^{18}O$ relative to the Dolomite I and I' by approximately 2.5 to 3.4‰ (Fig. 4.21). The post-dolomite blocky and columnar calcites are isotopically significantly lighter than Dolomites I, I', and II.

The corrected Sr isotope ratios of Dolomite I, I', and II also exhibit distinctly different values and ranges (Tables 4.3 and 4.4). The 12 samples of Dolomite I analyzed for $^{87}Sr/^{86}Sr$ are tightly clustered from 0.708866 to 0.708915, with a mean of 0.708896. This range of 0.000049 is only 2.5 times larger than the ±0.00002 analytical uncertainty. The 4 samples of Dolomite I' analyzed are also tightly clustered and yield a range of more radiogenic values from 0.708981 to 0.709030, with a mean of 0.708988. This range of 0.000049 is identical to that of Dolomite I. In contrast, the 5 Dolomite II samples analyzed in this study exhibit a slightly more radiogenic range of $^{87}Sr/^{86}Sr$ values from 0.708954 to 0.709088, which have a mean of 0.709002. While the Dolomite II mean value is approximately equivalent to that of Dolomite I', the Dolomite II range of 0.0001 is twice the spread observed in Dolomite I and I' (Table 4.4).

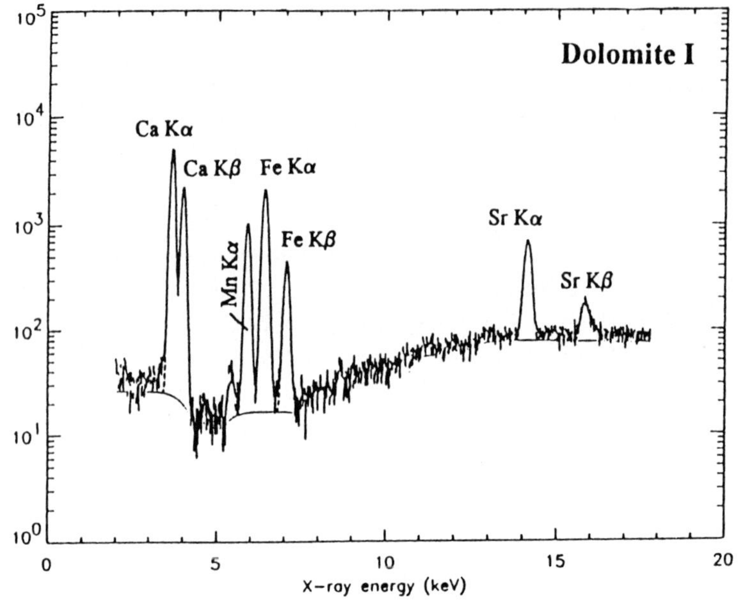
Dolomite I
Ca Kα
Ca Kβ
Fe Kα
Mn Kα
Fe Kβ
Sr Kα
Sr Kβ
X–ray energy (keV)

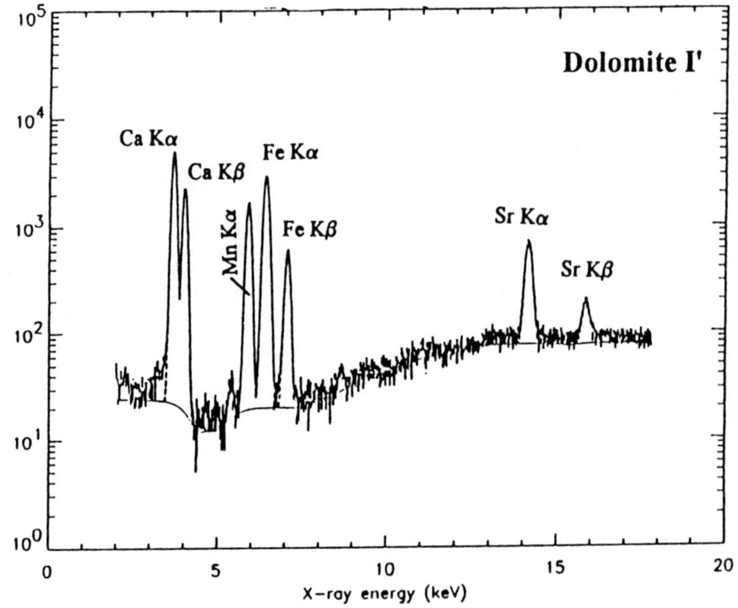
Dolomite I'
Ca Kα
Ca Kβ
Fe Kα
Mn Kα
Fe Kβ
Sr Kα
Sr Kβ
X–ray energy (keV)

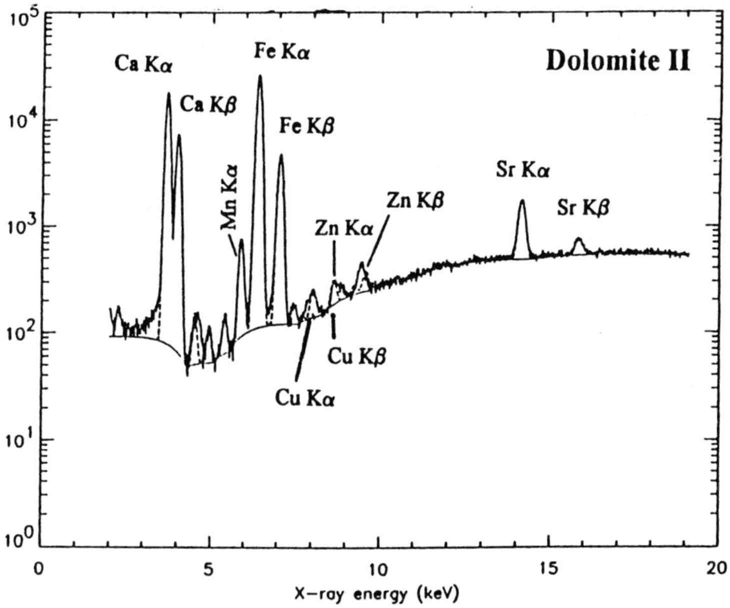

FIGURE 4.22. Synchrotron X-ray microprobe (SXRFMA) spectrum of Dolomites I, I', and II. *Top.* Dolomite I spectrum with Zn and Cu below detection (< 2 ppm). Sample number 2/14/89-II1 from Subunit 1 Lithofacies 1 grainstones at St. Michielsberg. *Middle.* Dolomite I' spectrum with Zn and Cu below detection (< 2 ppm). Sample number 2/25/89-III1 from Subunit 1 Lithofacies 1 grainstones at Seru Prekstul. *Bottom.* Dolomite II spectrum with Zn and Cu at concentrations of approximately 20 ppm. Sample number 9/16/90-Ia from Subunit 2 Lithofacies 7 grainstones at Seru Spreit.

Spatial Distributions in Dolomite Geochemistry

Several significant regional and stratigraphic trends have been observed in Dolomite I, I', and II geochemistry.

Geographic Distributions: Dolomites I, I', and II are approximately equivalent in their $\delta^{18}O$ values, with Dolomite I only slightly lighter than Dolomites I' and II (Table 4.4). Dolomite I, I', and II are also approximately equivalent in their Ca, Sr, Na, F, Cl, and SO_4 abundances. Therefore, no regional trends in $\delta^{18}O$, Ca, Sr, Na, F, Cl, and SO_4 were observed (Figs. 4.23 to 4.26). Although there are significant differences between the $^{87}Sr/^{86}Sr$ of Dolomite I and I', the small magnitude of Sr isotope variability (0.000049) with respect to the ana-

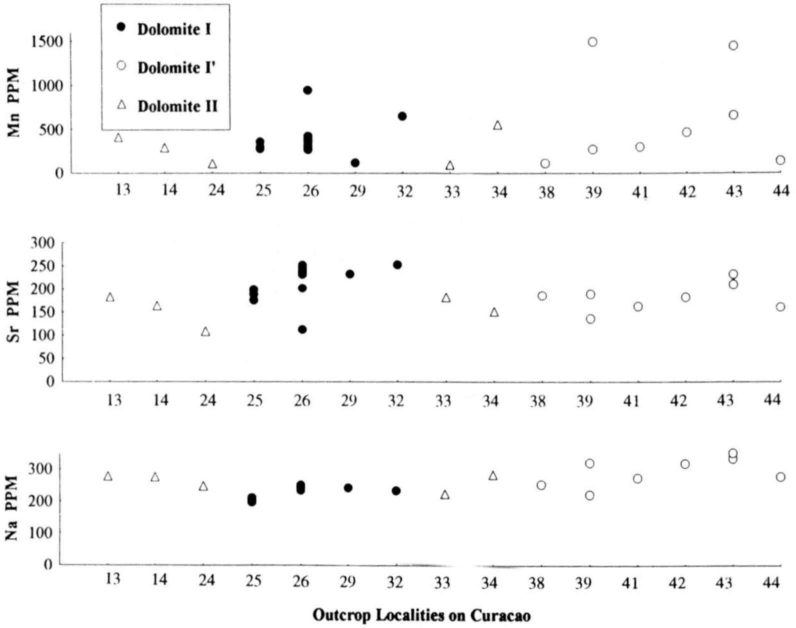

FIGURE 4.23. Regional variations in the Mn, Sr, and Na compositions of Dolomite I, I', and II. Sample locality numbers refer to those listed in Table 4.1 and Figure 4.5, and are as follows: 13 = Seru Kabaye; 14 = Seru Kabritu; 24 = Seru Spreit; 25 = St. Michielsberg; 26 = Seru Blandan; 29 = Veerisberg; 32 = Seru Domi; 33 = St. Annabaai; 34 = Fort Nassau; 38 = Seru Mansinga; 39 = Kabrietenberg; 41 = Tafelberg; 42 = Seru Mansinga; 43 = Seru Prekstul; 44 = Duivelsklip. Each data point represents a single analysis, and multiple analyses have been done on selected localities.

lytical uncertainty (± 0.00002) within each dolomite again prevented the recognition of consistent regional trends (Fig. 4.25). And while Dolomite II has a more significant spread in $^{87}Sr/^{86}Sr$ (0.0001), no regional trends were observed (Fig. 4.25). The only significant regional trend in dolomite geochemistry was observed in the $\delta^{13}C$ compositions of Dolomite I. Values decrease toward the northwest over a lateral distance of approximately 10 km a high of −2.55‰ at Veerisberg to significantly depleted lows of −5.40‰ at Seru Blandan and −6.79‰ at St. Michielsberg (Fig. 4.26).

Stratigraphic Distributions: Updip-downdip geochemical transects were evaluated within Dolomite I stratiform dolomite bodies at Seru Blandan. Pervasively

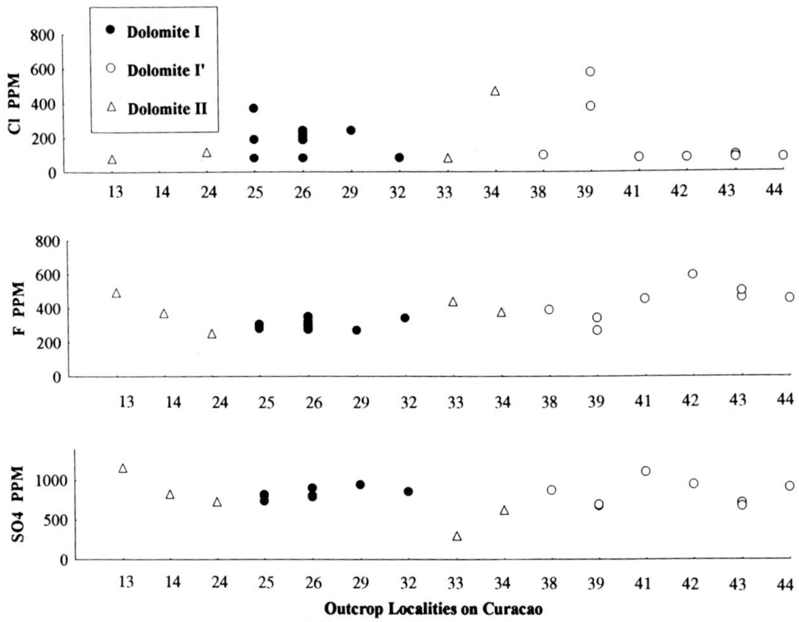

FIGURE 4.24. Regional variations in the Cl, F, and SO_4 compositions of Dolomite I, I', and II. Sample locality numbers refer to those listed in Table 4.1 and Figure 4.5. Each data point represents a single analysis, and multiple analyses have been done on selected localities.

dolomitized limestone samples from within the stratiform dolomite bodies were analyzed as well as partially dolomitized limestone samples comprising the downdip dolomite-limestone contact. The results of this reconnaissance analysis suggests that there is a significant downdip variation in $\delta^{18}O$ and $\delta^{13}C$ signatures. Conversely, Sr, Na, Cl, F, SO_4, Mn and Fe abundances and Sr isotope ratios in the dolomites are either invariant or exhibit erratic or random distributions and do not show consistent trends.

MODERN CURAÇAO GROUNDWATERS

The extent to which ancient waters may have interacted with the Curaçao basement is an important component in reconstructions of the fluids that precipitated Dolomites I, I', and II. Therefore, the geochemistry and hydrol-

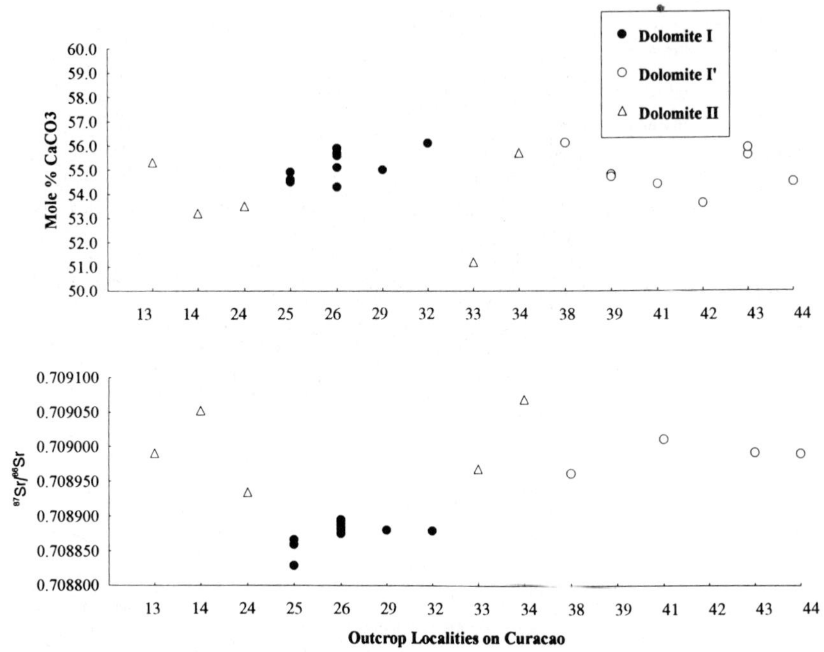

FIGURE 4.25. Regional variations in the mole percent $CaCO_3$ and $^{87}Sr/^{86}Sr$ compositions of Dolomite I, I', and II. Sample locality numbers refer to those listed in Table 4.1 and Figure 4.5. Each data point represents a single analysis, and multiple analyses have been done on selected localities.

ogy of the present-day groundwater system on Curaçao and Bonaire has been evaluated with data from a series of unpublished reports to the Island Government of Curaçao in the late 1970's. The groundwaters on Curaçao were analyzed by Rowbottom (1978) and Abtmaier (1979), and Bonaire groundwaters were analyzed by Rowbottom and Winkel (1979). These reports were made available for use in this study by Mr. C.W. Winkel of the Hydrology Division of the Department of Agriculture and Animal Husbandry of the Island Government of Curaçao. In addition, Cheatham (unpublished) presents Sr isotope analyses of modern Bonaire groundwaters.

The Curaçao groundwater report by Abtmaier (1979) is based on analyses of 2054 groundwater wells that penetrate the basement lithologies of the Curaçao Lava Formation, Knip Group, Midden Curaçao Formation, and the Quaternary Limestone Terraces. Abtmaier (1979) measured water level,

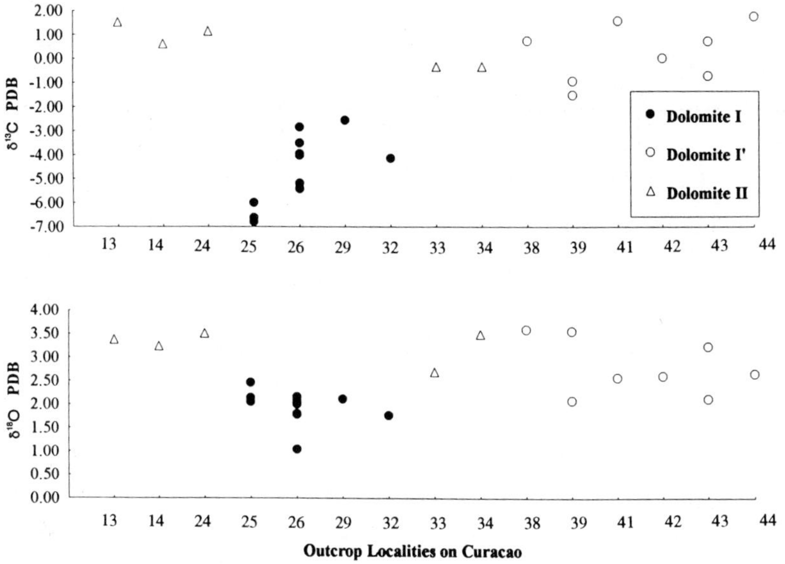

FIGURE 4.26. Regional variations in the $\delta^{18}O$ versus $\delta^{13}C$ compositions of Dolomite I, I', and II. Sample locality numbers refer to those listed in Table 4.1 and Figure 4.5. Each data point represents a single analysis, and multiple analyses have been done on selected localities.

pump-down time, conductivity, transmissivity, and water composition from 257 of these localities, and concluded that Curaçao is composed of a complex sequence of at least 15 drainage areas. Each area is hydrologically isolated from the others and has a unique mixture of basement lithologies and bedding geometries. The water chemistry analyses and conductivity measurements used for the present study from Abtmaier (1979) are exclusively from the southeastern half of Curaçao.

Geochemistry of Curaçao Lava Formation Groundwaters

The geochemical compositions of the groundwaters within the Curaçao Lava Formation chosen from southeastern Curaçao are presented in Table 4.5. In general, groundwater conductivity shows systematic increases toward each coastline due to increased mixing with Caribbean seawater. However, there is a local anomaly in conductivity approximately 2 km to the north of Tafelberg mountain (Locality 41 in Fig. 4.5). Ten strategically located water wells

A.

Well No.	pH	Ca PPM	Mg PPM	Na PPM	SO4 PPM	HCO3 PPM	Cl PPM	% Seawater	Calcite SI	Ca-Rich Dolomite SI
1	7.5	116	46	119	38	412	210	1.2	0.58	0.7
2	7.2	364	203	512	250	385	1593	9.0	0.51	0.68
3	8	404	218	956	480	197	2343	13.2	1.02	1.70
4	7.6	370	291	2146	432	217	3990	22.5	0.55	0.90
5	6.9	234	186	267	134	339	1108	6.3	-0.01	-0.21
6	7.3	88	63	149	34	464	249	1.4	0.29	0.35
7	7.6	110	64	177	53	395	375	2.1	0.61	0.91
8	8	212	71	368	130	488	732	4.1	1.30	2.08
9	7.7	202	118	207	115	351	792	4.5	0.85	1.39
10	7.4	60	78	168	38	346	318	1.8	0.10	0.19

B.

Well No.	pH	Ca PPM	Mg PPM	Na PPM	SO4 PPM	HCO3 PPM	Cl PPM	% Seawater	Calcite SI	Ca-Rich Dolomite SI
10	7.7	118	11	211	53	283	394	2.22	0.63	0.27
11	7.7	128	36	239	72	295	504	2.84	0.65	0.72
12	7.8	500	167	1800	720	290	2698	15.23	1.02	1.50
13	7.2	164	27	476	124	419	749	4.23	0.32	-0.14
14	7.6	114	10	216	105	409	292	1.65	0.66	0.29
15	7.4	452	284	989	320	300	2527	14.26	0.63	0.98
16	8.3	24	11	630	124	607	567	3.20	0.77	1.15
17	7.4	94	20	184	43	351	271	1.53	0.33	0.04
18	6.8	100	24	156	29	295	227	1.28	-0.40	-1.40
19	7.3	92	5	144	34	324	171	0.97	0.21	-0.78
20	7.1	110	20	138	34	390	191	1.08	0.11	-0.51
21	7	240	139	1603	360	331	2556	14.43	-0.19	-0.36
22	7.3	40	28	196	38	336	237	1.34	-0.15	-0.54
23	7.4	172	104	1152	316	285	2073	11.70	0.26	0.22
24	7.5	70	78	175	72	449	277	1.56	0.37	0.67
25	6.8	622	644	3427	982	446	7171	40.47	0.10	0.09
26	7.5	728	642	1152	310	275	9031	50.97	0.81	1.44
27	7.4	84	28	250	62	202	437	2.47	0.02	-0.48

TABLE 4.5. Modern Curaçao groundwater compositions from Abtmaier (1979). Saturation indices calculated with REACT in collaboration with Ron Stoessel. Estimates of percent seawater were calculated from Cl abundances. Locations of these wells are shown in Fouke (1993). A. Chemical compositions of modern Curaçao groundwaters within the Curaçao Lava Formation. B. Chemical compositions of modern Curaçao groundwaters within the Quaternary Limestone Terraces.

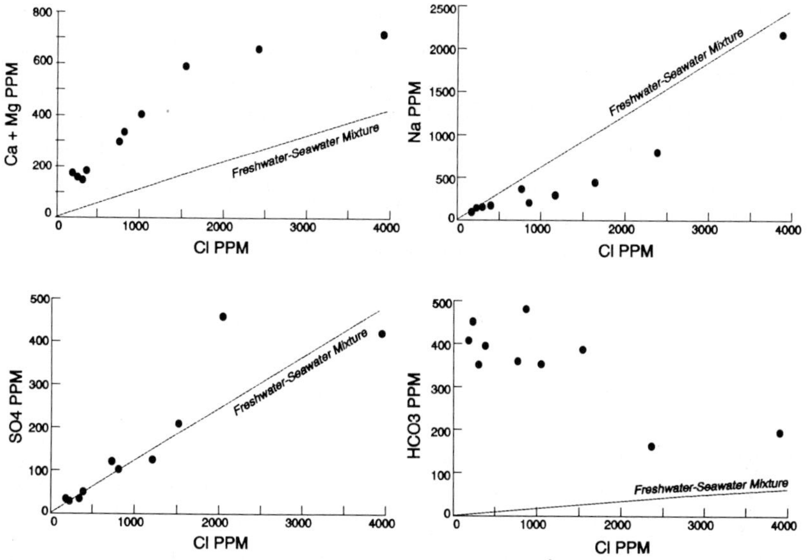

FIGURE 4.27. Cross-plots of Ca, Mg, Na, SO_4, and HCO_3 vs. Cl in modern Curaçao groundwaters. Data are presented in Table 4.5. In each cross-plot, the conservative behavior of Cl is used to evaluate whether the groundwater compositions are the result of seawater-freshwater mixing or water-rock interactions.

were analyzed in the region of this conductivity anomaly to evaluate whether the groundwater chemistries have changed significantly since being analyzed by Abtmaier (1979). Therefore temperature, pH, and alkalinity were measured in each of these wells in October of 1989. This evaluation yielded consistently good agreement with the data of Abtmaier (1979), and suggests that the general composition of Curaçao groundwaters in this region of the island have not changed drastically in the past 10 years.

In the following evaluation of the well water chemistries from Abtmaier (1979), Cl is used as a conservative estimate of seawater-freshwater mixing and is cross-plotted against Ca plus Mg, Na, SO_4 and HCO_3 (Fig. 4.27). The deviation of data points away from a freshwater-seawater mixing line of these crossplots indicates that the chemical modifications of these groundwaters can not be explained by mixing with seawater. Groundwaters exhibit significant enrichments in the abundance of Ca, Mg, and HCO_3 with increasing proximity toward the high conductivity anomaly to the north of Tafelberg.

Conversely, SO_4 remains conservative and parallels the mixing line, while Na exhibits significant depletion. Therefore some process in addition to mixing with seawater is selectively enriching Ca, Mg, and HCO_3 in the fluids, while depleting Na and leaving SO_4 relatively unaffected. Cheatham (unpublished) has analyzed Curaçao Lava Formation groundwaters on Bonaire for their Sr abundances and $^{87}Sr/^{86}Sr$ ratios (Table 4.6). The results reveal that the Bonaire groundwater water contains high Sr abundances of 0.11 to 3.06 ppm, and low $^{87}Sr/^{86}Sr$ values of 0.70614 to 0.70857.

Modelling of Fluid Saturation States and Mixtures

The Curaçao groundwater chemistries from Abtmaier (1979) have been modelled using the program REACT in collaboration with R. Stoessel at the University of New Orleans. REACT computes equilibrium assemblages at a given pressure or temperature in systems involving an aqueous fluid with or without mineral and gas phases (Stoessel *et al.*, 1989). This modelling has provided thermodynamic estimates of the saturation states for the Curaçao groundwaters with respect to calcite and dolomite. It has also permitted an evaluation of how these saturation states may evolve during mixing with seawater. In these calculations, the saturation index (SI) is defined as:

$$SI = \log (IAP / Ksp)$$

where IAP equals the ion activity product of the reaction and Ksp equals the thermodynamic equilibrium product of the reaction. An SI greater than 0 suggests that the solution is oversaturated with respect to the phase of interest, while an SI less than 0 suggests undersaturation for that phase. The Ksp dolomite for Ca-rich dolomites calculated by Stoessel *et al.* (1989) has been used in these calculations. The abundance of SO_4, and presumed absence of H_2S in the well waters suggest that sulphate reduction is not prominent in the Curaçao groundwater environments, and it would therefore not affect the fluid-saturation indices of the carbonate minerals (Stoessel, 1992).

Results of the REACT modelling suggest that the Curaçao Lava Formation groundwaters and most of the Quaternary Limestone Terrace groundwaters are supersaturated with respect to calcite and Ca-rich dolomite (Fig. 4.28) due to their high HCO_3 abundances. Curaçao groundwaters are therefore consistent with meteoric groundwaters in other parts of the Caribbean,

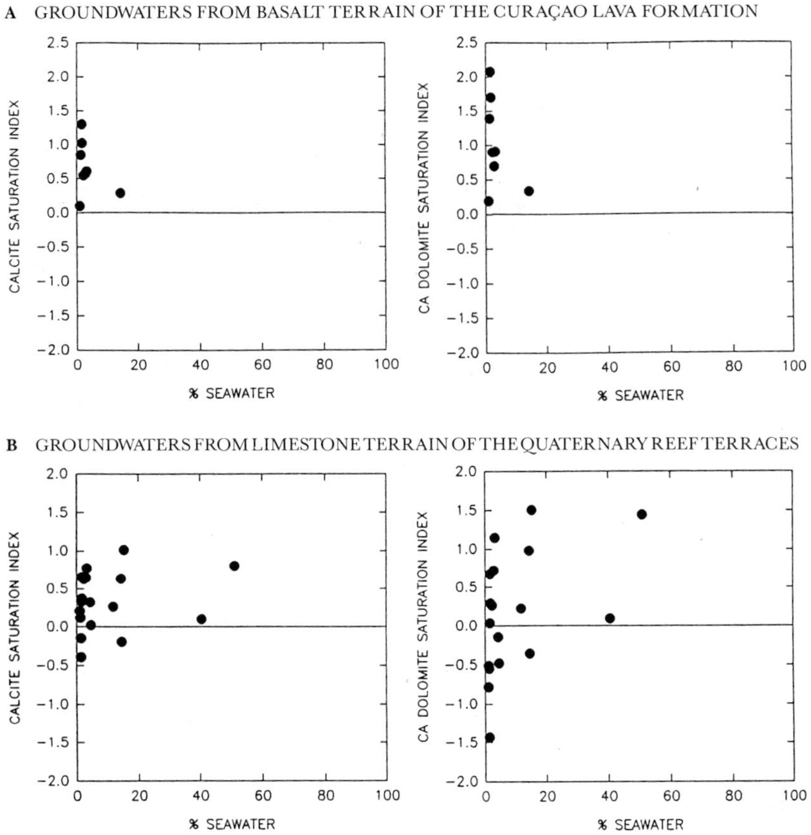

FIGURE 4.28. Saturation indices (SI) at 25°C and 1 atm for groundwaters from basaltic and carbonate terrains in southeastern Curaçao. Wells 1 through 10 are in the Curaçao Lava Formation, while wells 10 through 20 penetrate the Quaternary Limestone Terraces. Data used is from Abtmaier (1979) and is presented in Table 4.5. A. Saturation indices of Curaçao Lava Formation groundwaters with respect to calcite and Ca-rich dolomite. B. Saturation indices of Quaternary Limestone Terrace groundwaters with respect to calcite and Ca-rich dolomite.

which are also often oversaturated with respect to these phases (Plummer 1975; Ward and Halley, 1985; Stoessel *et al.*, 1989; Table 4.6) and *all* extents of mixing with seawater will also be oversaturated with respect to dolomite (Fig. 4.29). Therefore, no window of calcite undersaturation is achieved during the early to intermediate stages of fluid mixing, contrary to what is suggested by the Dorag model (see review in Hardie, 1987). This type of a seawater-Curaçao

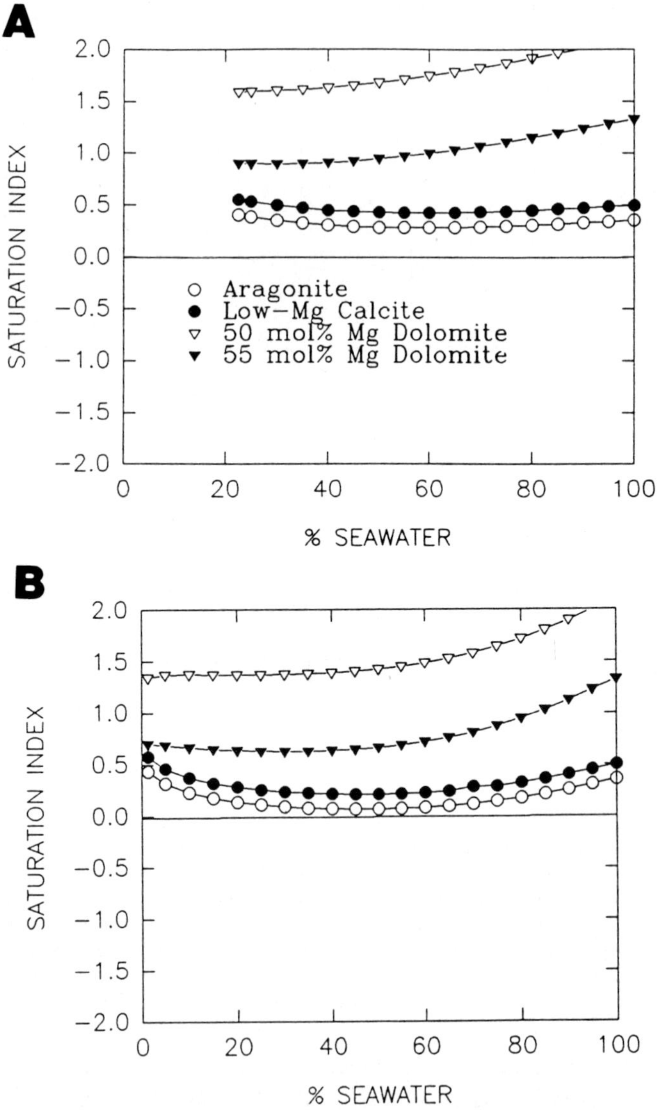

FIGURE 4.29. Saturation indices (SI) at 25°C and 1 atm for mixtures of selected Curaçao Lava Formation groundwaters with seawater. A. Mixing curves computed using well number 1 and Caribbean seawater from Stoessel *et al.* (1989). B. Mixing curves computed using well number 3 and Caribbean seawater from Stoessel *et al.* (1989). Data and well number designations are in Table 4.5.

Modern Near-Surface Diagenetic Fluids	Ca ppm	Mg ppm	Sr ppm	Mn ppm	Fe ppm	HCO3 ppm	$\delta^{18}O$ o/oo SMOW	$\delta^{13}C$ o/oo PDB	$^{87}Sr/^{86}Sr$	Reported Range
1. Fresh Groundwater	40	9	< 1	<<< 1	<< 1	< 300	< -2	< 0	(variable)	(averaged)
2. Curacao Groundwater	120	50	0.1	*nr*	*nr*	197	*nr*	*nr*	0.70614	(minimum)
(in basaltic terrain)	380	300	3.1	*nr*	*nr*	488	*nr*	*nr*	0.70857	(maximum)
3. Curacao Groundwater	40	5	*nr*	*nr*	*nr*	202	*nr*	*nr*	*nr*	(minimum)
(in carbonate terrain)	728	642	*nr*	*nr*	*nr*	450	*nr*	*nr*	*nr*	(maximum)
4. Bermuda Groundwater	0	0	0	*nr*	*nr*	5	*nr*	*nr*	*nr*	(minimum)
(in carbonate terrain)	377	1182	6.6	*nr*	*nr*	591	*nr*	*nr*	*nr*	(maximum)
5. Yucatan Groundwater	240	751	3	*nr*	*nr*	120	-0.1	-3.4	*nr*	(minimum)
(in carbonate terrain)	410	1283	8	*nr*	*nr*	250	-3.5	-11.1	*nr*	(maximum)
6. Caribbean Seawater	411	1290	8	0.0002	0.002	142	0	0	0.70917	(averaged)
7. Seawater-Derived	80	1373	*nr*	*nr*	*nr*	104	0.7	*nr*	*nr*	(minimum)
Bonaire Brines	1002	26006	*nr*	*nr*	*nr*	329	2.1	*nr*	*nr*	(maximum)
8. Hypersaline Brine	> 4000	> 580	> 5	< 1	< 1	> 100	> 0	< 15	*nr*	(averaged)

nr = not reported.

TABLE 4.6. The elemental and isotopic compositions of natural near-surface waters. 1. Fresh groundwaters from Budd (1984), Anderson and Arthur (1983), and Goldstein and Jacobson (1987). 2 and 3. Curaçao groundwaters from Abtmaier (1979) and Cheatham (unpub.). 4. Bermuda groundwater from Plummer *et al.* (1976). 5. Yucatan groundwaters from Stoessel *et al.* (1990). 6. Caribbean seawater from Drever (1982) and Hodell *et al.* (1991). 7. Seawater-derived brines on Bonaire from Deffeyes *et al.* (1964) and Lloyd (1966). 8. Hypersaline brine compositions from Carpenter and Miller (1969), McCaffrey *et al.* (1987), and Leaver (1985).

groundwater mixture could create dolomitizing fluids with unique geochemical compositions (i.e. low $^{87}Sr/^{86}Sr$ and transition-metal rich) that could be reflected in the solid-phase compositions of the dolomites that they precipitate.

Interpretation of the Origin of the Curaçao Lava Formation Groundwaters

The low Cl, Ca, Mg, Na, and SO_4, and high HCO_3 abundances of the anomalous Curaçao groundwaters with respect to seawater suggest that they are modified meteoric waters derived from rainfall, and not seawater-derived fluids. While the conservative behavior of SO_4 (Fig. 4.27) implies the mixing of up to 18% seawater, the nonconservative enrichment of Ca, Mg, and HCO_3 (Fig. 4.27) suggests that neither fluid mixing or evaporation (i.e. Bein *et al.*,

1991) is responsible for the observed geochemical alteration of the Curaçao groundwaters. The most geologically plausible cause of these anomalous groundwater compositions is water-rock interaction between meteoric fluids and the basalts comprising the Curaçao Lava Formation. These types of low-temperature reactions between basalt and freshwater are expected to produce Ca, Mg, and HCO_3 enrichments (Freeze and Cherry, 1979). In addition, the high abundances of less radiogenic Sr observed on Bonaire (Cheatham, unpublished) could only have been derived from interaction with the Cretaceous basement in this non-burial island setting. The south-eastern region of Curaçao, where the conductivity anomalies occur, is unique in that it contains (Fig. 4.30): 1) fault systems within the basement (Klaver, 1987); 2) MgO-rich Curaçao Lava Formation basalts and picrites (Klaver, 1987); and 3) the 90 and 100 mgal free-air gravity anomalies (Lagaay, 1969).

The quantitative modelling of fluid interaction with Curaçao Lava Formation basalts, as well as back-calculations of original solid-phase compositions from the groundwater compositions, are beyond the scope of the present study. However, low abundances of less radiogenic Sr (Sr=44 to 246 ppm, $^{87}Sr/^{86}Sr=0.70324$ to 0.70449; G.T. Klaver pers. comm.) have been observed in fresh samples of the tholeiitic basalts comprising the Curaçao Lava Formation. These low Sr abundances suggest that it would not be possible to attain the 3 ppm concentrations of dissolved Sr observed in groundwaters on Bonaire during open-system fluid-rock interaction. A potentially more viable process by which to source Sr to the groundwater is from water-rock interaction with basalts that have been weathered, such as the modern 3 to 18 m-thick regolith that has been formed atop the Curaçao Lava Formation basalts (Abtmaier, 1979). Sr within these weathered basalts would presumably be more easily leached than from a fresh basalt. Therefore semi-closed system water-rock interaction with the weathered basalts, coupled with evaporitic effects, could theoretically be sufficient to enrich the abundance of Sr within the fluids to the levels observed by Cheatham (unpublished) on Bonaire. This type of hydrologic environment is consistent with the observations of C.W. Winkel (pers. comm.), who has documented fluctuations in the meteoric water table of several meters through the course of a year as a result of seasonal recharge and evaporation cycles.

In conclusion, the results of these analyses have documented that modern Curaçao groundwaters can locally contain high abundances of less radio-

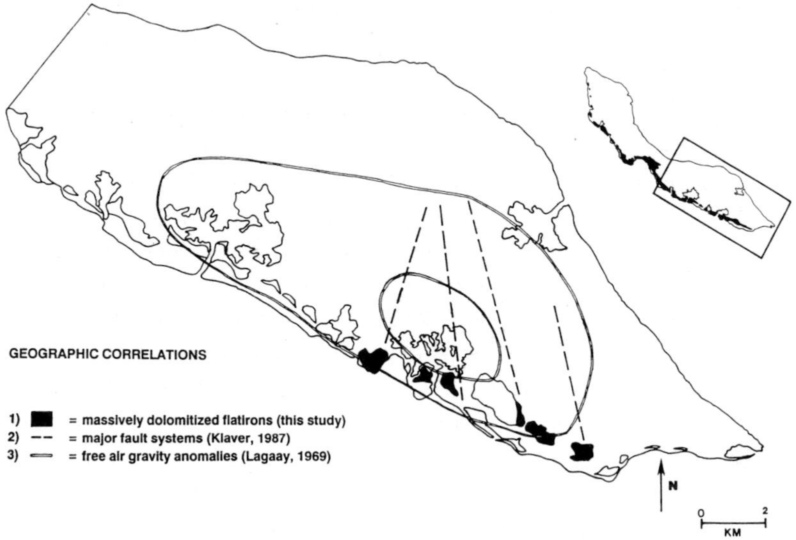

FIGURE 4.30. Geographic correlations on southeastern Curaçao of geologic and geochemical anomalies. The southeastern portion of Curaçao contains: 1) extensive pervasive dolomitization in the Seroe Domi Formation (present study); 2) major fault systems in the basement composed of MgO-rich basalts and picrites (Klaver, 1987); and 3) free-air gravity anomalies (Lagaay, 1969).

genic Sr derived from water-rock interaction of meteoric fluids with weathered basalts. The high Sr abundances and low $^{87}Sr/^{86}Sr$ in the groundwaters require that relatively small amounts of this water be mixed with seawater to significantly alter the Sr isotopic composition of the resulting mixed fluid.

Evaluation of Dolomite Recrystallization

A hierarchy of petrographic criteria has been compiled in this study in order to establish that Dolomite I, I', and II crystals are primary precipitates from solution that have not experienced recrystallization. Disproving that recrystallization has occurred permits the direct reconstruction of ancient diagenetic fluid compositions from solid-phase dolomite geochemistries. Recrystallization has been evaluated using the techniques of Ward and Reeder (1993) in which the orientation of larger-scale optically visible crystal growth features (overall morphology, sectoral boundaries, and concentric, sectoral, and intrasectoral zonations) are compared with the orientation of small-scale growth features resolvable at the scale of the TEM (microstructural zoning, growth dislocations, and modulated microstructures). Consistency in orientation of these macro- and microstructural criteria is evidence of primary precipitatation from solution (Ward and Reeder, 1993).

Microstructural TEM analyses of Dolomite I' are presented in Fouke and Reeder (1992). The orientations of pervasive fine-scale modulated microstructures and screw disslocation surfaces in Dolomite I' parallel the larger scale concentric CL zonations and cleavage form of the sector in which they occur (Fouke and Reeder, 1992). It is presumed that these concentric CL zonations would be destroyed or altered during recrystallization, therefore their presence is further indicative of primary crystallization (e.g. Reeder and Prosky, 1986; Cander *et al.*, 1988; Machel and Anderson, 1989; Meyers, 1991). Therefore, because of the consistency in orientation of these large and small scale features, it is interpreted that these dolomite crystals are primary precipitates that have not been recrystallized.

Reconstruction of Estimated Seawater Dolomite Compositions

In order to geochemically evaluate the dolomitization of the Seroe Domi Formation, it is essential to first establish a comparative geochemical standard reflecting the composition of contemporaneous seawater and the host limestones during each dolomitization event (e.g. Meyers and Lohmann, 1985; Banner *et al.*, 1988). In this section, the isotopic and elemental compositions of an estimated seawater dolomite (ESD) will be established with which to compare the geochemical compositions of Dolomite I, I', and II.

Oxygen and Carbon Isotopes

Analyses of Tertiary benthic foraminifer tests from continuous DSDP cores in the Atlantic suggest that the $\delta^{18}O$ of calcite precipitated from seawater varied from +1.0 to +3.0‰ PDB during the Miocene-Pliocene to +3.0 to +3.8‰ PDB during the Pleistocene. Analyses of pelagic foraminifers accompanying the benthic foraminifers in the DSDP cores suggest that the $\delta^{13}C$ of seawater also varied during the Neogene from approximately +0.5 to +2.5‰ PDB

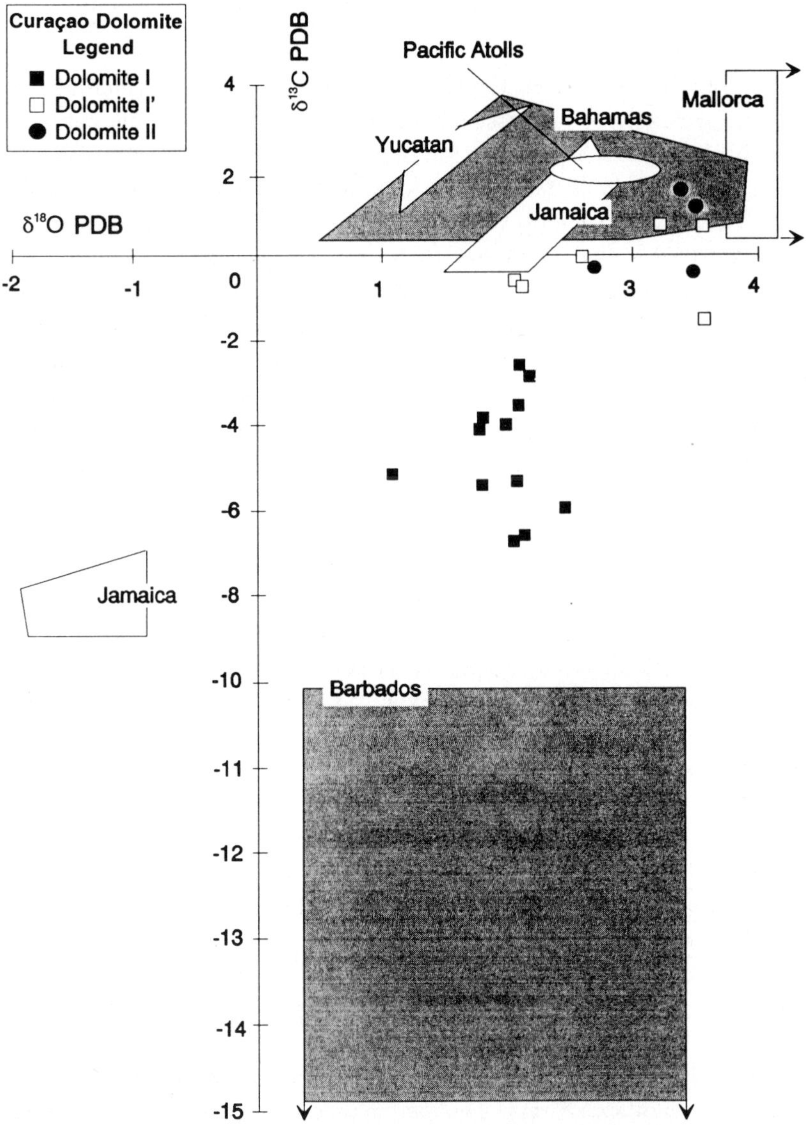

FIGURE 4.31. Comparative cross-plot of Dolomite I, I', and II $\delta^{18}O$ versus $\delta^{13}C$ with other Tertiary dolomites. Sources of data: Barbados from Humphrey (1988); Jamaica from Land (1973a, b); Yucatan from Ward and Halley (1985); Bahamas from Supko (1977) and Vahrenkamp (1988); Mallorca from Oswald (1992); Pacific Atolls from Saller (1984), Aharon *et al.* (1989), and Hein *et al.* (1992).

(Shackleton and Kennett, 1975; Shackleton and Cita, 1979; Miller *et al.*, 1987, 1988).

An intraformational $\delta^{18}O$ and $\delta^{13}C$ seawater calcite standard for the Seroe Domi Formation was not possible to establish because all available texturally pristine oyster, brachiopod, and foraminifer shell material was either significantly geochemically altered, or contained authigenic calcite cements. Therefore, the broad range suggested by the DSDP foraminifer analyses of +1.0 to +3.8‰ PDB $\delta^{18}O$ and +0.5 to +2.5‰ PDB $\delta^{13}C$ (Miller *et al.*, 1987) are accepted as an estimate of Neogene seawater calcite for the present study. An average empirically-derived dolomite-calcite $\delta^{18}O$ fractionation factor of 3 ±1‰ has been added to this range, while no fractionation factor is applied to the dolomite $\delta^{13}C$ (Land, 1980). This suggests an estimated seawater dolomite (ESD) composition of +3.0 to +5.8‰ $\delta^{18}O$ PDB and +0.5 to +2.5‰ $\delta^{13}C$ PDB (Table 4.4). These values are also consistent with Tertiary dolomites interpreted to have precipitated from seawater in other studies (Fig. 4.31).

Strontium Isotopes

Hodell *et al.* (1991) have developed a regression age model based on observed increases of from 0.7087 to 0.7092 in the $^{87}Sr/^{86}Sr$ of seawater since the Early Miocene from analyses of pelagic foraminifer tests in continuous DSDP and ODP cores. This $^{87}Sr/^{86}Sr$ age model suggests that an ESD precipitated during the Middle Miocene (synchronous with Subunit 1 deposition) should contain an $^{87}Sr/^{86}Sr$ of 0.70871 to 0.70886. Similarily, a Late Miocene to Pliocene ESD (synchronous with Subunit 2 deposition) should have a $^{87}Sr/^{86}Sr$ signature of 0.70886 to 0.70908, while a Pleistocene ESD should contain an $^{87}Sr/^{86}Sr$ of 0.70908 to 0.70918.

Trace Element Abundances

Estimated seawater dolomite (ESD) trace element abundances will be established in two different ways. The first method is to calculate trace element abundances using effective partition coefficients published in other studies. Because of the many uncertainties associated with the determination of partition coefficients, this approach provides only broad compositional estimates. The second approach is to use the composition of dolomites from other localities that are interpreted to be precipitated from seawater. It is therefore assumed that the overlap in ranges generated by these two methods will provide reasonable estimates of ESD trace element compositions.

Strontium: Strontium is incorporated into the calcite crystal lattice by substitution at Ca structural sites (Pingitore *et al.*, 1992), therefore the fluid Sr/Ca ratio determines the amount of Sr that is ultimately incorporated in calcite. Several estimates have been made of the effective distribution coefficient (k^{Sr}dolomite) for dolomite, which relates the Sr/Ca ratio in solution to the ratio of the dolomite that it precipitates and is described by the equation:

$$(Sr/Ca)\text{dolomite} = k^{Sr}\text{dolomite} \times (Sr/Ca)\text{fluid}$$

Estimates of k^{Sr}dolomite for Ca-rich dolomites range from 0.025 to 0.14 (Kinsman, 1969; Jacobson and Udowski, 1976; Katz and Matthews, 1977; Baker and Burns, 1985). These Sr partitioning coefficients predict that an ESD precipitated directly from seawater (Sr/Ca=8 ppm/411 ppm=0.02; from Drever, 1982) should contain 108 to 657 ppm Sr.

Vahrenkamp and Swart (1990) have illustrated that Late Tertiary seawater dolomites from the Bahamas have a strong correlation between Sr composition and Ca enrichment, and propose a k^{Sr}dolomite of 0.118 for stoichiometric dolomite with a 0.0039 coefficient increase for each additional mole % $CaCO_3$. Calculations using this relationship with the stoichiometries observed in Dolomites I, I', and II predict that an ESD should contain 75 to 172 ppm Sr. The 165 to 184 ppm concentrations of Sr observed at Enewetak and other localities (Fig. 4.32) are consistent with the upper end of the range suggested with the Vahrenkamp and Swart (1990) stoichiometry calculations. This suggests an ESD composition of 75 to 184 ppm Sr (Table 4.4).

Sodium, Sulphate, Chlorine, and Fluorine: The concentration of Na has been used to interpret the salinity of diagenetic fluids in several dolomitization studies (e.g. Weber, 1964; Fritz and Katz, 1972; Land and Hoops, 1973; Veizer *et al.*, 1977; Sibley, 1980; Choquette and Steinen, 1980; Humphrey, 1988; Oswald, 1992; Staudt *et al.*, 1993). However, it is uncertain whether Na is incorporated in the dolomite lattice by major element substitution or interstitial incorporation, and the value of the Na partition coefficient itself has never been experimentally determined (Hoops and Land, 1973; Staudt *et al.*,

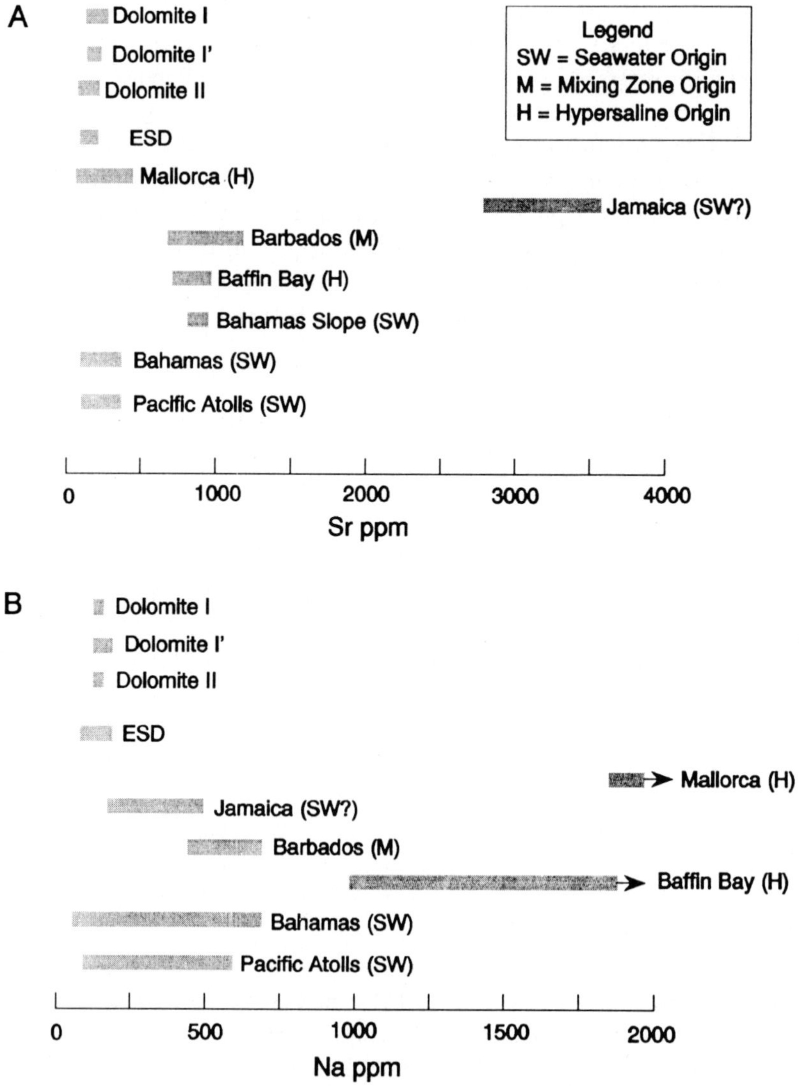

FIGURE 4.32. Comparison of Sr and Na abundances in Dolomites I, I', and II and other Tertiary dolomites. Sources of data: a) Barbados from Humphrey (1988); Jamaica from Land (1973a, b); Yucatan from Ward and Halley (1985); Bahamas from Supko (1977) and Vahrenkamp (1988); Bahamas slope from Mullins *et al.* (1985); Mallorca from Oswald (1992); Pacific Atolls from Saller (1984) and Aharon *et al.* (1989); Baffin Bay from Behrens and Land (1972).

1993). Miocene Enewetak seawater dolomites contain 290 to 375 ppm Na (Saller, 1984), Cretaceous Judea Group seawater dolomites contain 150 to 350 ppm Na (Sass and Bein, 1989), and Permian Lower Clear Fork Formation seawater dolomites contain 184 to 343 ppm Na (Staudt *et al.*, 1993; Fig. 4.32). Therefore, the combination of these studies suggest an ESD composition of 150 to 375 ppm Na (Table 4.4).

Sulfate is similar to Na, in that it is apparently readily incorporated into dolomite. However, little is known about whether it is interstitial, substitutional, or is sourced from sulphate-rich inclusions such as gypsum (Staudt *et al.*, 1993). The size of the SO_4 tetrahedron suggests that it substitutes for CO_3 in dolomite rather than being incorporated into interstitial sites, as is the mechanism observed in calcite (Takano *et al.*, 1980; Takano, 1985; Busenburg and Plummer, 1985). And as with Na and SO_4, it is unknown where Cl and F occurs in dolomite. Modern carbonate sediments of the Florida Keys contain bulk sediment F abundances of 600 to 800 ppm, and concentrations in aragonitic skeletal material that range from 30 to over 1600 ppm (Rude and Aller, 1991).

Recent analyses of the SO_4, Cl, and F composition of dolomites by Staudt *et al.* (1993) suggests that significant differences in these trace elements occur between dolomites interpreted to be precipitated from burial, seawater, mixed seawater, and hypersaline fluid environments (Figs. 4.33 and 4.34). Seawater to mixing zone dolomites were observed to contain 157 to 494 ppm Cl, and 232 to 2625 ppm SO_4, and from less than 80 to 1500 ppm F (Staudt *et al.*, 1993), which will be accepted as the ESD compositional ranges in the present study (Table 4.4).

Manganese and Iron: Experimental studies suggest that Mn and Fe substitute more readily for Mg than Ca in dolomite, and that over 80% of the Mn in dolomite resides in the Mg sites (Lumsden and Lloyd, 1988; Lumsden *et al.*, 1989). However, the concentration of Mn and Fe in solution is strongly dependent upon the reducing or oxidizing state of the solution (Stumm and Morgan, 1980). This redox sensitivity, in addition to the extremely large partition coefficients (much greater than 1; Veizer, 1983) and low abundances of 0.2 ppb Mn and 2 ppb Fe in normal oxygenated seawater (Drever, 1982), make the abundance of Mn and Fe in dolomite strongly dependent upon local environmental factors (e.g. flow rate and resulting water-rock ratios). Therefore, it is difficult to constrain adequately the calculation of

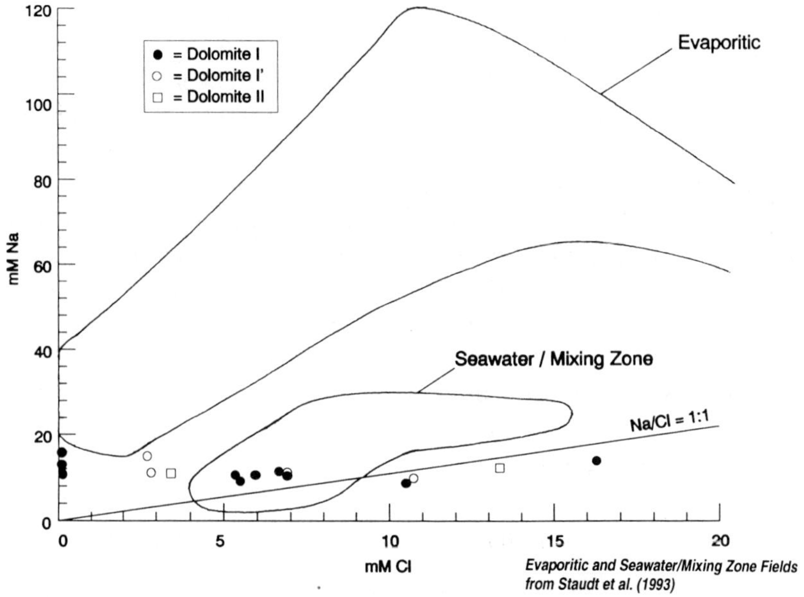

FIGURE 4.33. Cross-plot of Na versus Cl in Dolomites I, I', and II compared to dolomites interpreted to have precipitated from seawater, mixed seawater, hypersaline, and burial environments. Data fields from Staudt *et al.* (1993).

predicted abundances of Mn and Fe for seawater dolomite (Veizer, 1983). Because of these uncertainties, an ESD range of 16 to 57 ppm Fe and 19 to 22 ppm Mn will be used in this study (Table 4.4). These abundances are derived from analyses of the Enewetak seawater dolomites (Saller, 1984) and are consistent with the low abundances for a seawater dolomite predicted by Veizer (1983).

Zinc and Copper: It is uncertain whether Zn and Cu substitute for Ca or Mg in the dolomite crystal lattice (Tsuesue and Holland, 1966; Kitano *et al.*, 1980). Both Zn and Cu have large partition coefficients (5 to 10; Kitano *et al.*, 1968; 1980), and occur in extremely low abundances in seawater (0.5 ppb Cu and 2 ppb Zn; Drever, 1982). These constraints make the abundances of Zn and Cu in a seawater-derived dolomite sensitive to flow rate and resulting water-rock interaction ratios. Therefore their abundances in an ESD are difficult to predict, other than to suggest that they should be extremely low (Veizer, 1983).

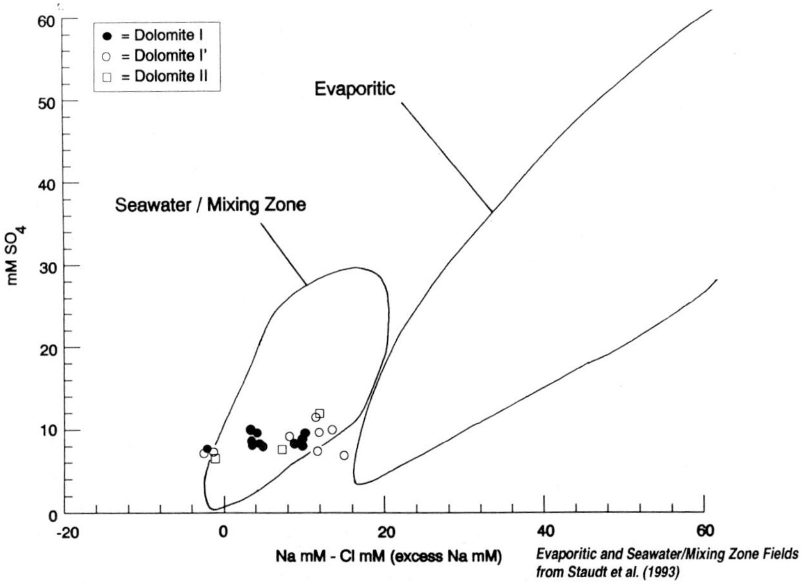

FIGURE 4.34. Cross-plot of SO_4 versus Na and Cl in Dolomites I, I', and II compared to dolomites interpreted to have precipitated from seawater, mixed seawater, hypersaline, and burial environments. Data fields from Staudt *et al.* (1993).

Studies of subsurface seawater and/or mixing zone dolomites from the Nieu and Aitutaki Pacific atolls have documented abundances of 5 to 500 ppm Cu and 90 to 300 ppm Zn in dolomite whole-rock analyses (Aharon *et al.*, 1987; Hein *et al.*, 1992). However, these variable abundances are interpreted to be from interstitially precipitated low-temperature iron hydroxide laminae in the dolomitized limestones, and not from the dolomites themselves. In a study of Carboniferous dolomites, whole-rock analyses revealed concentrations of up to 9 ppm Zn and 5 ppm Cu, which were similarly attributed to interstitial and siliciclastic contamination (Barber, 1974). Conversely, whole-rock analyses of hypersaline dolomites from modern salinas on the Pacific atolls of Minamidaitojima, Yoron, Bora Bora, Kikai, and Moorea, which may be free of interstitial iron hydroxides, have yielded average compositions of 1.6 ppm Cu and 0.4 ppm Zn (Aizawa and Akaiwa, 1988). Therefore, the combination of the low abundances in seawater and analyses of these hypersaline Pacific atoll dolomites, suggest that ESD Zn and Cu concentrations should be less than 2 ppm (Table 4.4).

Composition of Dolomite I, I', and II Dolomitizing Fluids

Dolomites I, I', and II have similar compositions with respect to their stoichiometry and abundances of Sr, Na, Cl, F, SO_4, Fe, and Mn, all of which are consistent with precipitation from seawater or a seawater-dominated diagenetic fluid. Therefore, this suite of geochemical characteristics will be interpreted simultaneously prior to the individual interpretations of $\delta^{18}O$, $\delta^{13}C$, $^{87}Sr/^{86}Sr$, Zn and Cu for Dolomite I, I', and II.

Stoichiometry

Dolomites I, I', and II are Ca-rich, ranging from 52.1 to 56.1 ±1.5 mole% $CaCO_3$ (Table 4.4). Compositional variability between crystals in a single thin-section reaches 2 mole%, which coupled with the 1.3% analytical uncertainty prevents distinction between Dolomites I, I', and II samples. Therefore, no geographic, stratigraphic, or geochemical covariations in Dolomites I, I', and II stoichiometry could be discerned. Several studies have attempted to use the Ca/Mg composition of dolomites to interpret the Ca/Mg of the fluids from which they precipitated, but no consensus exists regarding a direct relationship between fluid and solid ratios (Goldsmith and Graf, 1958; Katz, 1973; Folk and Land, 1975; Lumsden and Chimahusky, 1980; Sass and Bein, 1989; Wogelius *et al.*, 1992). Sass and Bein (1989) suggest that the $CaCO_3$ content of dolomite may increase with decreases in salinity and Ca/Mg of the dolomitizing fluid. This is based on comparisons of dolomite interpreted to be of seawater and hypersaline origins. If correct, this would imply that Ca-rich dolomites such as Dolomite I, I', and II are more consistent with precipitation from nonevaporitic seawater-like fluids rather than a hypersaline brine.

Strontium, Sodium, Fluorine, Chlorine, and Sulfate

The trace elements Sr, Na, F, Cl, and SO_4 exhibit consistently low abundances in Dolomites I, I', and II (Table 4.4). These compositions are similar to those predicted for the ESD and are therefore suggestive of precipitation from a seawater-dominated fluid. Dolomite I, I', and II Sr concentrations are similar to those reported from low-temperature Neogene dolomites in other localities interpreted to be of seawater and/or mixing zone origins, while being equivalent in concentration to significantly lower than dolomites interpreted to have precipitated from hypersaline fluids (Fig. 4.32). However, the relatively narrow range in Dolomite I, I', and II stoichiometry and Sr abundances prevents quantitative correlation with the Bahamas dolomite k^{Sr}dolomite function described by Vahrenkamp and Swart (1990). Low Sr abundances comparable to those of Dolomite I, I', and II have often been attributed to recrystallization during early dolomitization processes (Veizer, 1983; Land, 1985; Carballo *et al.*, 1987). However, no evidence of recrystallization was observed in the Seroe Domi Formation dolomites and the low Sr abundances in Dolomites I, I', and II are interpreted to be primary geochemical signatures.

The abundance of Na in Dolomite I, I', and II is consistent with the predicted ESD concentrations (Table 4.4) as well as with concentrations in dolomites interpreted to be of seawater and mixed seawater origin (Fig. 4.32). Conversely they are significantly lower than concentrations in dolomites interpreted to have been precipitated from hypersaline fluids (Fig. 4.32). In addition, the concentrations of excess Na (defined as Na abundance minus Cl abundance) versus SO_4 are equivalent to those reported from low-temperature dolomites in other localities that are interpreted to be of seawater and/or mixing zone origins (Fig. 4.34).

Petrographic features, in conjunction with the analysis of Cl and SO_4 abundances in Dolomite I, I', and II, were used to evaluate the possibility that Na was sourced either from contamination during sample preparation, or from solid and fluid inclusions within the dolomites. Dolomitized Seroe Domi Formation limestones are intermixed with up to 50% basalt-derived sands. However, the basalts comprising these sands originally contain low Na abundances (Klaver, 1987), and leaching experiments (Fouke, 1993) illustrate that the basaltic sands yield insignificant amounts of Na and Si during dissolution. Fluid inclusions are extremely rare in Dolomite I, I', and II, and no halite or anhydrite solid inclusions were observed under optical (plane-light, cathodoluminesence, fluoresence, SEM) and TEM (Reeder, pers. comm.) analyses. However, SO_4 abundances were too low (1159 to less than 300 ppm) to exclude the presence of gypsum based on mass balance calculations. Dolomites I, I', and II also have consistently low abundances of Cl (576 to less than 90 ppm), and have molar Na/Cl ratios ranging from 0.8 to 6.5 with 95% of the ratios greater than 1 (Fig. 4.33). Na/Cl ratios consistently less than or equal to 1 would be expected if all of the Na had been derived from human sweat during sample preparation, fluid inclusions, or solid halite or anhydrite inclusions (Sass and Bein, 1988; Staudt *et al.*, 1993). Therefore, all of the Na measured in Dolomite I, I', and II is interpreted to have been sourced from within the dolomite and not from contamination or inclusions. The low but consistent abundance of SO_4 in Dolomites I, I', and II is compatible with concentrations observed in other dolomites interpreted to have precipitated from normal oxygenated seawater (Fig. 4.34), and suggests that the sulfate inhibition dolomitization model may not be applicable (Baker and Kastner, 1981; Gunatilaka *et al.*, 1987; Hardie, 1987). The relatively low variability of SO_4, Na, and Cl in Dolomites I, I', and II prevent meaningful interpretations of geographic, stratigraphic, and geochemical covariations (Figs. 4.23 and 4.24).

Fluoride concentrations in Dolomite I, I', and II are low, falling within a narrow range from 254 to 592 ppm. The low abundance of F in seawater (1 ppm; Rude and Aller, 1991) coupled with the low abundance of F in Dolomite I, I' and II with respect to dolomites interpreted to be of evaporitic origin (Staudt *et al.*, 1993) further suggest derivation from a seawater-like fluid. However, as with the other trace elements, the low variability of F in Dolomite I, I', and II prevented covariation analyses with other parameters.

Manganese and Iron

The concentrations of Mn and Fe in Dolomite I, I', and II are orders of magnitude larger than those of the ESD (Table 4.4). However, the interpretation of Mn and Fe is complicated due to the complexities involved with the fluid chemical behavior of Mn and Fe and their incorporation controls in dolomite (e.g. Machel and Burton, 1991). The cathodoluminesence observed in Dolomite I, I', and II is interpreted to be the result of Mn^{2+} activation and variable quenching due to varying Fe^{2+} contents, while the concentric CL-defined zoning is suggestive of multiple temporal changes in fluid composition during crystallization resulting from variations such as changes in fluid concentrations, pH, Eh, growth rate, and temperature (Sommer, 1972; Meyers, 1974; Pierson, 1981; Hemming *et al.*, 1989; Barnaby and Rimstidt, 1989; Morse and Bender, 1990; Machel and Burton, 1991). In addition, the sectoral zoning observed in Dolomite I' documents consistent differences in composition between the 1014 and 1011 growth sectors, which illustrates that surface structure influences dolomite composition as well as fluid parameters (Fouke and Reeder, 1992).

The 2000 to 3000 ppm concentrations of Mn and Fe in Dolomite I, I', and II (Fouke and Reeder, 1992) are up to 6 times larger than those observed in pre- and post-dolomite calcite cements. This indicates that Mn and Fe in the dolomite are fluid dominated, and were not sourced from the previously stabilized host limestones. The solubility of Mn and Fe is much greater under reducing conditions, while Fe is removed from solution much more quickly than Mn upon oxidation (Stumm and Morgan, 1981). Therefore, the high abundance of both Mn and Fe in Dolomite I, I', and II implies reducing conditions in the dolomitizing fluids (e.g. Humphrey, 1988; Meyers, 1991). Reducing freshwaters commonly have high Mn and Fe abundances on the order of 10 to 100 ppm respectively (Veizer, 1983), and measurements from Bonaire groundwaters suggest Fe concentrations of 0.4 ppm (B. Lyons, pers. comm.). Therefore, a mixture of these groundwaters with seawater may have created dolomitizing fluids with sufficently high Mn and Fe abundances to attain the trace element abundances observed in Dolomite I, I', and II. Increased extents of water-rock interaction could also enrich Mn and Fe due to their extremely large ($\gg 1$) partition coefficients (Veizer, 1983).

Potential Sources of Sr to the Dolomitizing Fluids

Prerequisites for applying $^{87}Sr/^{86}Sr$ dating techniques to Dolomite I, I', and II are to demonstrate that: 1) *all* of the Sr within each dolomite was derived solely from seawater; and 2) the Sr has not been subsequently mobilized during diagenesis. The consistency in large and small petrographic fabrics suggests that the $^{87}Sr/^{86}Sr$ of Dolomite I, I', and II reflect the composition of the original diagenetic fluids from which they precipitated. This permits the direct comparison of Dolomite I, I', and II $^{87}Sr/^{86}Sr$ with the $^{87}Sr/^{86}Sr$ of the host Seroe Domi Formation limestones in order to evaluate the sources of Sr in the dolomitizing fluids.

The Seroe Domi Formation occurs in an island setting containing four possible sources of Sr, three of which have distinctly different $^{87}Sr/^{86}Sr$ compositions and Sr abundances (Fig. 4.35). The first possible source of Sr are the Cretaceous basalts of the Curaçao Lava Formation and associated basalt-derived siliciclastics of the Midden Curaçao and Knip Group formations comprising the Curaçao basement (Beets, 1972). These contain Sr

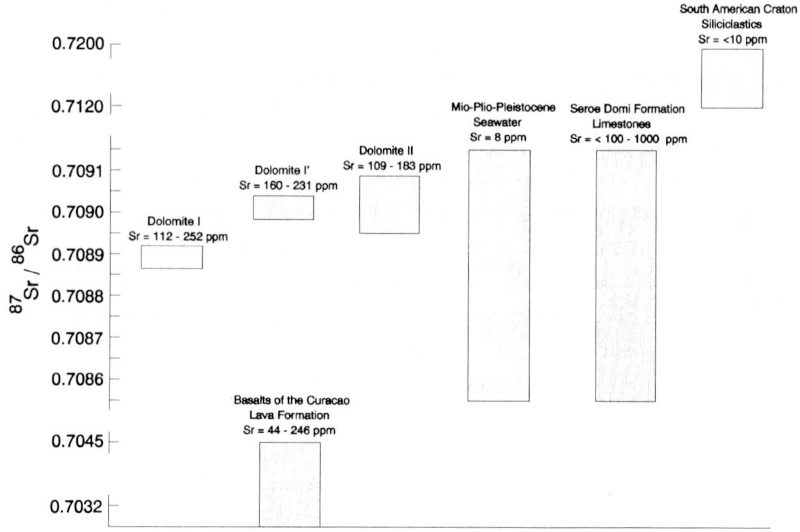

FIGURE 4.35. Comparison of $^{87}Sr/^{86}Sr$ and Sr abundances from various sources on Curaçao with Dolomite I, I', and II.

abundances of 44 to 246 ppm and low $^{87}Sr/^{86}Sr$ values of 0.703239 to 0.704488 (Klaver, pers. comm.; Fig. 4.35).

The second possible source of Sr are the Seroe Domi Formation and Quaternary Terrace Limestones on Curaçao (Fig. 4.35). Extensive pre-dolomite meteoric stabilization suggests that Subunit 1 limestones were incapable of providing a local source of Sr to Dolomite I and I' fluids. Conversely, it is possible that Dolomite II derived some Sr from either the Subunit 2 limestones or the Quaternary Terrace Limestones. However, the small 0.00005 difference in their original $^{87}Sr/^{86}Sr$ of these two sources, coupled with the low abundance of Sr in meteoric fluids (Table 4.6), suggest that deciphering the mixing of these two sources would be difficult given the ± 0.00002 analytical uncertainty in this study.

A third possible source of Sr to the Seroe Domi Formation dolomites is contemporaneous seawater at the time of dolomitization (Fig. 4.35). The oldest rocks in which Dolomite I occurs are Subunit 1, therefore any Mio-Plio-Pleistocene and Recent seawater $^{87}Sr/^{86}Sr$ equivalent to and younger than these limestones might have been involved in dolomitization. The possible range of $^{87}Sr/^{86}Sr$ from the Middle Miocene Subunit 1 limestones is 0.70871 to 0.70916 (after Hodell *et al.,* 1991).

And finally, a fourth possible source of Sr to the Seroe Domi Formation dolomites are quartz sands derived from the South American craton observed in the Knip Group and Midden Curaçao Formation (Beets, 1972; Fig. 4.35). Siliceous dusts in the Caribbean atmosphere derived from both the Sahara and South America constitute a subset of this category, with loads that can range from 2 to 20 ug/m^3 of air/day (Prospero *et al.,* 1970). However, although these sands and dusts would have distinct radiogenic signatures of 0.716 ±0.004 (Faure and Powell, 1972; Faure, 1986), their extremely small volumetric proportions and low Sr abundances suggest that they were not a viable source of Sr to the dolomitizing fluids.

Dolomite I, I', and II Chronology

Several lines of geologic and geochemical evidence will be used with respect to the sources of Sr on Curaçao, to demonstrate that the Sr in Dolomites I and I' was derived from seawater. Conversely, Dolomite II yields a wide range in $^{87}Sr/^{86}Sr$, the highest of which will be shown to represent seawater Sr, while the lowest of which represents an mixture between Sr derived from the basement and seawater Sr.

Timing of Dolomite I Precipitation

Dolomite I contains $^{87}Sr/^{86}Sr$ values of 0.708866 to 0.708915, which range from being equivalent to more radiogenic than the 0.70879 $^{87}Sr/^{86}Sr$ of the host Middle Miocene Subunit 1 limestones. Therefore, the Sr in Dolomite I is suggestive of being derived from seawater due to: 1) the consistent, tightly clustered $^{87}Sr/^{86}Sr$ signatures of Dolomite I'; 2) the equivalent to more radiogenic Sr of Dolomite I than the host limestone; and 3) seawater-like trace element compositions. Regression age modelling of Dolomite I $^{87}Sr/^{86}Sr$ suggests a late Middle Miocene to early Late Miocene age of 9.2 to 12.1 ±1.4 Ma (after Hodell *et al.,* 1991; Table 4.7; Fig. 4.36). The restriction of Dolomite I to Middle Miocene Subunit 1 Lithofacies 1, 2, and 3 limestones suggests that dolomitization took place sometime prior to deposition of the overlying Late Miocene to Pliocene Subunit 2 Lithofacies 6 lime-

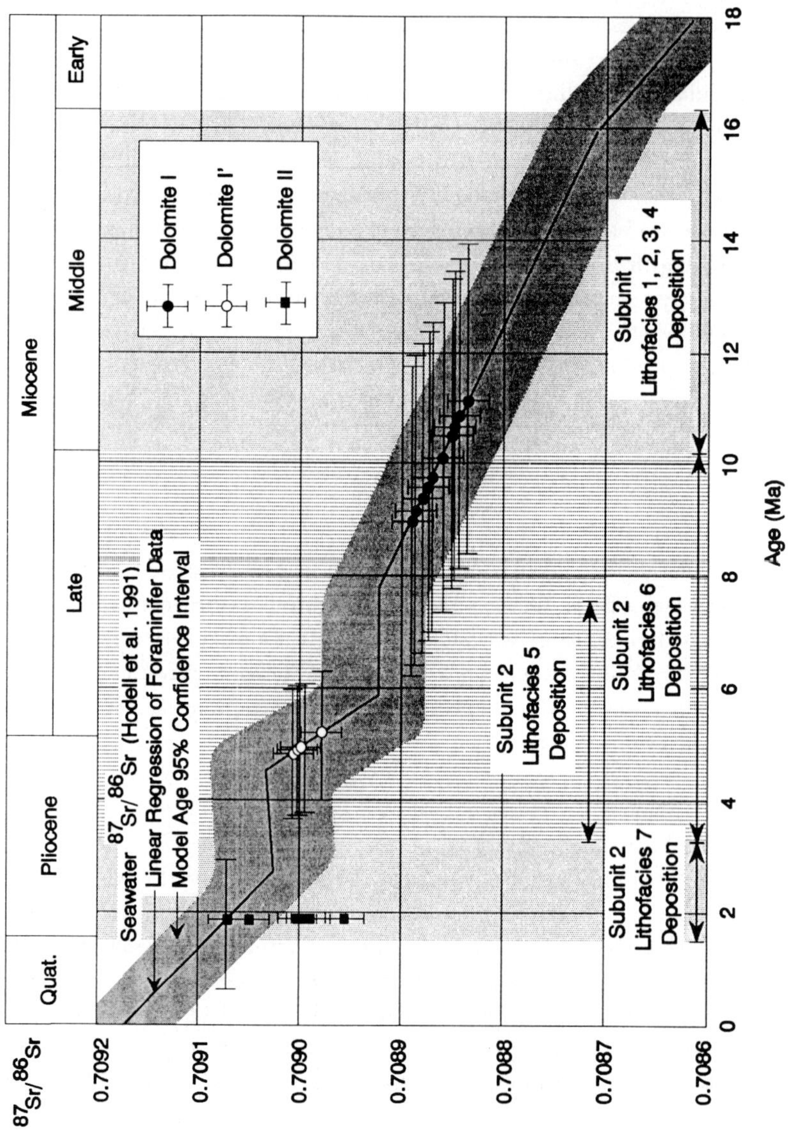

FIGURE 4.36. Cross-plot of Dolomites I, I', and II $^{87}Sr/^{86}Sr$ versus their model ages and the age of the host Seroe Domi Formation limestones. Based on regression age model developed from DSDP foraminifer analyses by Hodell *et al.* (1991).

			$^{87}Sr/^{86}Sr$		Hodell et al. (1991)	
Sample Locality	Sample Number	Sample Composition	Measured Ratio	In-Run Uncertainty	Interval Number	Model Age
St. Michielsberg	2/14/89-II4	Dolomite I	0.708846	0.000021	N-V	10.6
		duplicate	0.708812	0.000070	N-V	12.1
		average	0.708829		N-V	11.3
St. Michielsberg	2/14/89-II5	Dolomite I	0.708866	0.000046	N-V	9.8
St. Michielsberg	2/14/89-II9	Dolomite I	0.708848	0.000033	N-V	10.5
		duplicate	0.708870	0.000029	N-V	9.6
		average	0.708859		N-V	10.1
Seroe Blandan	2/13/89-I3N	Dolomite I	0.708875	0.000021	N-V	9.4
Seroe Blandan	2/21/89-I3N	Dolomite I	0.708875	0.000036	N-V	9.4
Seroe Blandan	2/21/89-II5	Dolomite I	0.708886	0.000018	N-V	8.9
Seroe Blandan	3/29/89-II5	Dolomite I	0.708891	0.000018	N-V	8.7
Seroe Blandan	3/29/89-II7	Dolomite I	0.708885	0.000012	N-V	9.0
Seroe Blandan	3/30/89-V2	Dolomite I	0.708895	0.000014	N-V	8.5
Seroe Blandan	3/30/89-IV3	Dolomite I	0.708880	0.000018	N-V	9.2
Veerisberg	5/23/88-II7	Dolomite I	0.708880	0.000012	N-V	9.2
Fort Nassau	5/27/88-III1	Dolomite I	0.708879	0.000018	N-V	9.2
Average			0.708875		N-V	9.4
Caracasbaai	6/25/88-V3	Dolomite I'	0.708961	0.000029	N-III	5.2
Tafelberg	7/1/88-V7	Dolomite I'	0.709010	0.000015	N-III	4.8
Seroe Prekstul	2/25/89-III1	Dolomite I'	0.708991	0.000050	N-III	4.9
Duivelsklip	2/26/89-I3	Dolomite I'	0.708989	0.000018	N-III	4.9
Average			0.708988		N-III	5.0
Seru Largu	5/22/90-II7	Dolomite II	0.708990	0.000009	N-III	4.9
Seroe Spreit	9/16/90-IIA	Dolomite II	0.708934	0.000012	N-III	5.4
St. Annabaai	4/5/89-I	Dolomite II	0.708968	0.000010	N-III	5.1
Seru Kabaye	5/11/90-IA	Dolomite II	0.709052	0.000011	N-I	2.1
Fort Nassau	5/30/88-I7	Dolomite II	0.709068	0.000012	N-I	1.9
Average			0.709002		N-III	4.8

TABLE 4.7. $^{87}Sr/^{86}Sr$ geochronology model age calculations for Dolomites I, I', and II. Ages determined with regression age model presented by Hodell *et al.* (1991). A constant of 0.00002 was subtracted from the Sr isotope data in this study for normalization with the analyses of Hodell *et al.* (1991).

stones. This stratigraphic evidence directly supports the late Middle Miocene to early Late Miocene age for Dolomite I suggested by $^{87}Sr/^{86}Sr$ geochronology.

Timing of Dolomite I' Precipitation

Dolomite I' occurs within Subunit 1 Lithofacies 1, 2, and 4 limestones, as well as within Subunit 2 Lithofacies 5. The Dolomite I' $^{87}Sr/^{86}Sr$ values of 0.708981 to 0.709030 are significantly more radiogenic than the 0.70879 to 70893 $^{87}Sr/^{86}Sr$ of the host Middle Miocene Subunit 1 limestones. However,

Dolomite I' $^{87}Sr/^{86}Sr$ values are equivalent to those of the host Subunit 2 Lithofacies 5. Therefore, the Sr in Dolomite I' is suggestive of being derived from seawater due to: 1).the consistent, tightly clustered $^{87}Sr/^{86}Sr$ signatures of Dolomite I; 2) the equivalent to more radiogenic Sr of Dolomite I' than the host limestone; and 3) seawater-like trace element compositions, Regression age modelling of Dolomite I' $^{87}Sr/^{86}Sr$ suggests a a late Late Miocene to Pliocene age of 9.2 to 12.1 $\pm$1.4 Ma (after Hodell *et al.,* 1991; Table 4.7; Fig. 4.36). The presence of Dolomite I' in Subunit 2 Lithofacies 5 limestones suggests that the age of precipitation can further be constrained to an interval from late Late Miocene to Early Pliocene.

Timing of Dolomite II Precipitation

Unlike Dolomites I and I', Dolomite II exhibits relatively wide variations in $^{87}Sr/^{86}Sr$ of 0.00013, with values that range from 0.708954 to 0.709088. Therefore, the upper half of the Dolomite II $^{87}Sr/^{86}Sr$ range is equivalent to the $^{87}Sr/^{86}Sr$ of Lithofacies 7, while the lower half of this range is significantly less radiogenic than the host limestone (Fig. 4.36). This wide spread in Dolomite II $^{87}Sr/^{86}Sr$, in conjunction with it being less radiogenic than the host limestones and the seawater-like geochemical compositions, suggests that Dolomite II $^{87}Sr/^{86}Sr$ represents a mixture between seawater Sr and a less radiogenic Sr source. Therefore, as a minimum estimate of the age of precipitation, regression age modelling of the most-radiogenic Dolomite II $^{87}Sr/^{86}Sr$ suggests precipitation from an Early Pleistocene mixed Sr fluid at 1.9 $\pm$0.6 Ma (after Hodell *et al.,* 1991; Table 4.7; Fig. 4.36). This minimum age is consistent with the Late Pliocene to Early Pleistocene depositional age of the host Lithofacies 7 limestones. Because all of the older Subunit 1 and 2 limestones were stabilized prior to the precipitation of Dolomite II, the only remaining sources of less radiogenic $^{87}Sr/^{86}Sr$ were fluids containing basalt-derived Sr (Fig. 4.35). The modern Curaçao Lava Formation groundwaters provide evidence that meteoric fluids containing high abundances less radiogenic Sr do occur on Curaçao (Table 4.6).

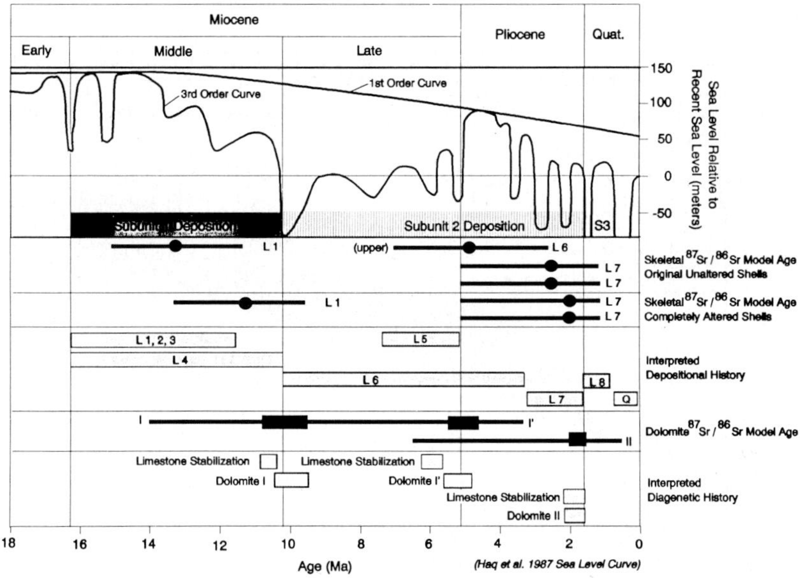

FIGURE 4.37. Summary of stratigraphic and geochemical evidence used to constrain the ages of Seroe Domi Formation deposition, meteoric diagenesis, and dolomitization. Events are correlated with global third-order sealevel changes presented by Haq *et al.* (1987).

Dolomitization History of the Seroe Domi Formation

The above geologic and geochemical evaluations of Dolomites I, I', and II suggest that fluids similar in composition to seawater may have been responsible for the dolomitization of the Seroe Domi Formation. However, a seawater-like fluid origin for Dolomites I, I', and II must be able to explain: 1) the distinct geographic and stratigraphic distributions of Dolomite I, I', and II within Subunit 1 and 2 limestones; 2) the occurrence of multiple dolomitization events; 3) the differences between Dolomite I, I', and II $\delta^{18}O$, $\delta^{13}C$, and $^{87}Sr/^{86}Sr$; and 4) the enrichment of Zn and Cu in Dolomite II samples containing less radiogenic $^{87}Sr/^{86}Sr$. In the following discussion, each of these considerations will be evaluated with respect to evidence from: 1) the host limestone paragenetic sequence; 2) the tectonically and eustatically controlled elevation of relative sea level; and 3) the spatial distributions of dolomite geochemical compositions (Figs. 4.37 and 4.38).

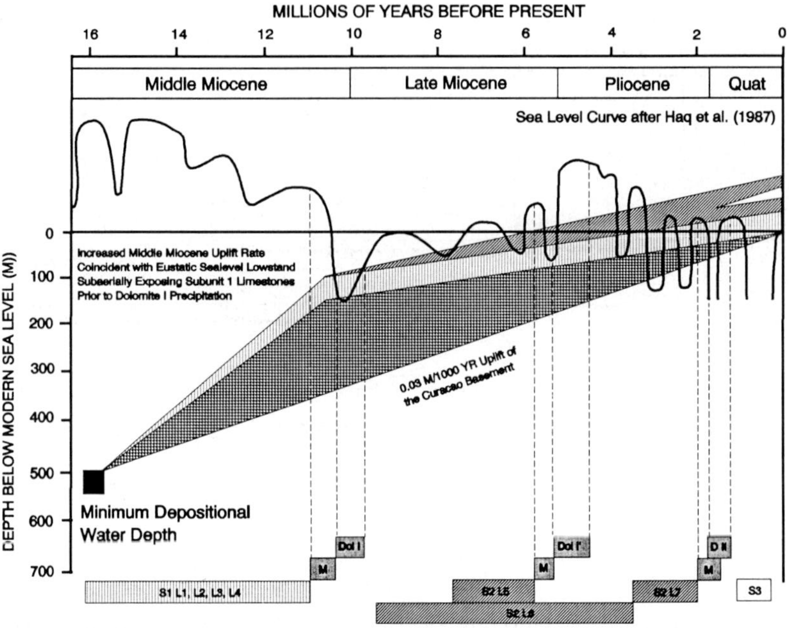

FIGURE 4.38. Interpreted depositional, diagenetic, and uplift history of the Seroe Domi Formation. Events are correlated with global third-order sealevel changes presented by Haq *et al.* (1987). The depositional sequences are represented by the following: Subunit I Lithofacies 1, 2, 3, and 4 = S1 L1 L2 L3 L4; Subunit 2 Lithofacies 5 and 6 = S2 L5 L6; Subunit 2 Lithofacies 7 = S1 L7; and Subunit 3 Lithofacies 8 = S3 L8. Each episode of meteoric diagenesis (M) and Dolomite I (Dol I), Dolomite I' (Dol I') and Dolomite II (D II) precipitation correlates with a significant drop in third-order global sealevel. The overall uplift rate of 0.03 m/1000 yr was probably exceeded during the Middle Miocene, which acted in conjunction with the eustatic drawdown to subaerially expose Subunit 1 limestones prior to Dolomite I precipitation.

Pre-Dolomite I and I' Diagenesis

The intense leaching of skeletal grains, replacement of oyster shells, and precipitation of low-Mg calcite (LMC) cements occurred in Subunit 1 limestones prior to Dolomite I and I' precipitation. These features are similar to cements and dissolution fabrics that occur during meteoric diagenesis and suggest that Subunit 1 limestones were subaerially exposed prior to Dolomite I and I' precipitation (e.g. Land, 1973b; Matthews, 1974; James and Choquette, 1983; Budd, 1984; Saller and Moore, 1989). Similar leaching and cementation fabrics have been observed in deep marine environments (Saller and Koepnick, 1990). However the simultaneous depletion in O and C isotopes and Sr, and simulataneous enrichment in Mn and Fe, are most consistent with meteoric fluid environments and ensuing water-rock interactions (e.g. Meyers and Lohmann, 1985; Lohmann, 1987).

Differences in $\delta^{18}O$ between the skeletal material and pre-Dolomite I and I' calcite cement suggests that a two-stage meteoric stabilization process took place in Subunit 1 limestones prior to dolomitization. In the first stage, LMC oyster shells were replaced by LMC, which suggest a value for Middle Miocene meteoric calcite of approximately −2.25‰ PDB $\delta^{18}O$ (Fig. 4.39). The second stage of stabilization represented by the pre-dolomite LMC bladed calcites suggests a value for late Middle Miocene meteoric calcite of approximately −5.4‰ PDB $\delta^{18}O$ just prior to dolomitization (Fig. 4.39).

The source of Sr in the meteoric fluids that altered the Subunit 1 oyster shells is difficult to quantitatively evaluate, and may have been a mixture of Sr derived from the host Subunit 1 carbonates as well as contemporaneous seawater. Therefore, a $^{87}Sr/^{86}Sr$ model age established from this altered shell material would be a *maximum* estimate of the age of stabilization, suggesting ages that are slightly too old due to the incorporation of the less radiogenic Subunit 1 limestone Sr (e.g. Vahrenkamp *et al.*, 1988). The modelling of the $^{87}Sr/^{86}Sr$ compositions in texturally unaltered, partially altered, and completely altered Subunit 1 oyster shells suggests that Subunit 1 limestone stabilization occurred no earlier than the late Middle Miocene to early Late Miocene at approximately 11.2 $\pm$1.6 Ma. This estimate is consistent with the paragenetic sequence, in that it suggests a maximum age for meteoric stabilization that predates the 9.4 to 11.3 $\pm$1.4 Ma Late Miocene $^{87}Sr/^{86}Sr$ model age of Dolomite I precipitation. A maximum age of late Middle Miocene also correlates well with third-order global sealevel draw downs of approximately −200 m associated with the onset of Antarctic glaciation (Haq *et al.*, 1987; Figs. 4.37 and 4.38). Two driving forces acted simultaneously to expose the limestones at this time, which include the tectonic uplift of the Curaçao horst and third-order eustatic sealevel lowstand (Fig. 4.38). Without a eustatic sealevel fall, the Subunit 1 lithologies would have remained at deep marine water depths (> 200 m) during the late Middle Miocene, precluding hydrologic contact with meteoric and/or mixed seawater-freshwater fluids (Fig. 4.38). Differential updip-downdip exposure of the Subunit 1 lime-

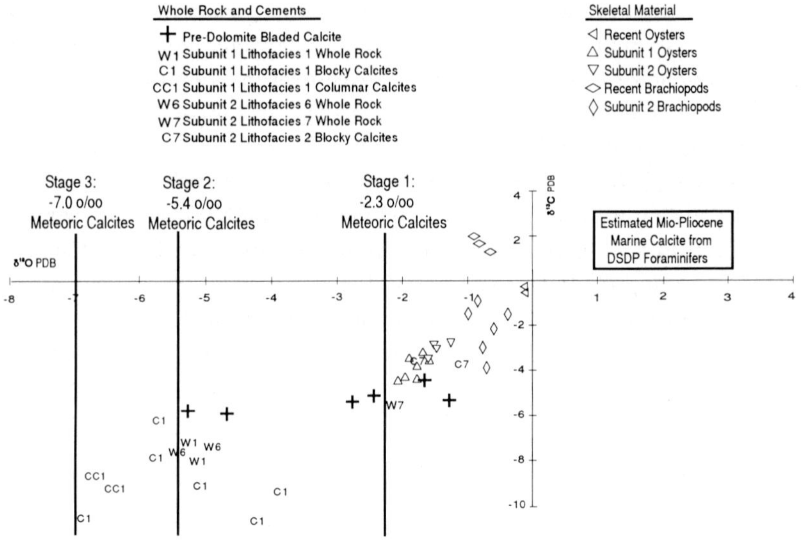

FIGURE 4.39. Crossplot of $\delta^{18}O$ and $\delta^{13}C$ compositions of Seroe Domi Formation calcite cements and skeletal material. At least three stages of meteoric calcitization are suggested by the shifting $\delta^{18}O$ composition of each group of calcite cements.

stones is suggested by the significantly greater abundance of pre-dolomite LMC bladed and blocky calcites in Lithofacies 3 allochthonous blocks than in surrounding Lithofacies 1 grainstones. These relationships imply that the updip lithologies from which Lithofacies 3 were derived may have periodically experienced more extensive meteoric diagenesis due to more frequent and/or prolonged subaerial exposure during the Middle Miocene.

Petrogenesis of Dolomite I

Dolomite I contains low abundances of Sr, high abundances of Fe and Mn, and consistent $^{87}Sr/^{86}Sr$ signatures. These analyses suggest that Dolomite I precipitated from reducing fluids similar in composition to seawater at 9.4 to 11.3 ± 1.4 Ma during the late Middle Miocene to early Late Miocene (Figs. 4.37 and 4.38). This implies that Dolomite I precipitation is slightly younger than the estimated Subunit 1 limestone stabilization age of 11.2 ± 1.6 Ma (Fig. 4.38), and is consistent with the relative sequence of leaching, calcitization, and dolomitization suggested by the paragenetic sequence

(Fig. 4.18). These ages for pre-dolomite meteoric diagenesis and Dolomite I precipitation correlate well with the third-order eustatic sea level lowstand and ensuing sea level rise at the Middle to Late Miocene transition (Haq *et al.*, 1988; Fig. 4.38).

Composition of the Dolomitizing Fluids: The $\delta^{18}O$, $\delta^{13}C$, and $^{87}Sr/^{86}Sr$ compositions of Dolomite I suggest that the dolomitizing fluids were composed of a mixture of seawater and meteoric groundwater. Dolomite I $\delta^{18}O$ versus $\delta^{13}C$ data correlate closely with modelled trajectories representing precipitation from a mixture of seawater and freshwater (modelled after Langmuir *et al.*, 1978), rather than trajectories representing fluid alteration during water-rock interaction (modelled after Banner and Hanson, 1990; Fig. 4.40; Table 4.3). The intermediate position of the Dolomite I $\delta^{18}O$ versus $\delta^{13}C$ data between the estimated seawater dolomite (ESD) and estimated freshwater dolomite (EFD) endmembers also imply that these seawater-freshwater mixtures contained as much as 50% freshwater (Fig. 4.40). Dolomite I samples that fall below the linear mixing line between the ESD and EFD total dissolved carbon (TDC) suggests that the freshwater had higher TDC abundances than the seawater with which it was mixed ($K \geqslant 0.15$; Fig. 4.40). The $\delta^{13}C$ compositions of two of the Seru Blandan samples and all three of the St. Michielsberg samples are isotopically lighter than the EFD $\delta^{13}C$ composition (Fig. 4.40). This suggests that the composition of the meteoric waters on Curaçao varied spatially and temporally, which is further suggested by the regional depletion trends in Dolomite I $\delta^{18}O$ and $\delta^{13}C$ toward the St. Michielsberg outcrop (Fig. 4.26).

Covariations between $^{87}Sr/^{86}Sr$ versus $\delta^{18}O$ and $\delta^{13}C$ could not be used to distinguish between fluid mixing and water-rock interaction due to the narrow range of Dolomite I $^{87}Sr/^{86}Sr$ compositions (Fig. 4.41). However, the close correlation of binary mixing curves with Dolomite I $^{87}Sr/^{86}Sr$ versus Sr and 1/Sr compositions also supports a mixed fluid origin (Fig. 4.42). Modelled hyperbolic asymptotes of this curve suggest that the original seawater had a composition of 0.708877, while the original freshwater contained 0.001 ppm Sr (Fig. 4.42). The radiogenic $^{87}Sr/^{86}Sr$ signatures of Dolomite I imply that the freshwaters that mixed with seawater were not directly analogous to groundwaters on modern Curaçao. Mass balance calculations suggest that the high abundance of Sr in modern Curaçao groundwaters would have significantly lowered the Sr isotopic signature of the resulting fluid at

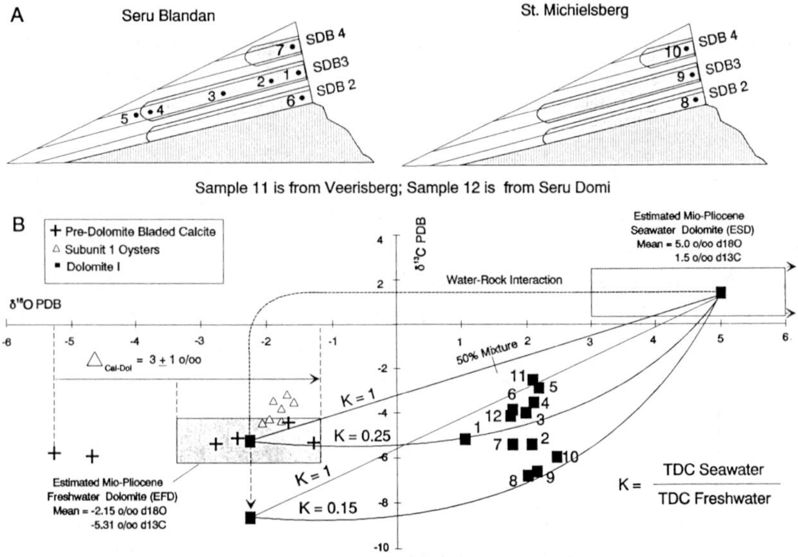

FIGURE 4.40. Evaluation of Dolomite I $\delta^{18}O$ versus $\delta^{13}C$ compositions with respect to dolomite precipitation from mixed seawater and freshwater. Sample numbers correspond to those presented in Table 4.3. A. Schematic depiction of the spatial distribution of Dolomite I samples, including: 1) Samples 1, 2, 3, 4, and 5 from an updip-downdip transect within stratiform dolomite body (SDB) 3 at Seru Blandan; 2) Samples 6, 1, and 7 from a vertical transect across SDB 2, 3, and 4 at Seru Blandan; 3) Samples 8, 9, and 10 from a vertical transect across SDB 2, 3, and 4 at St. Michielsberg; and 4) Samples 11 and 12 are reconnaissance samples from Veerisberg and Seru Domi respectively. B. Cross-plot of Dolomite I O and C isotopic compositions with respect to modelled seawater-freshwater fluid mixing trajectories. Estimated freshwater dolomite (EFD) compositions are reconstructed by assuming dolomite $\delta^{13}C$ equilibrium with freshwater and the addition of a 3 ±1‰ calcite-dolomite $\delta^{18}O$ fractionation factor to the most isotopically depleted pre-dolomite bladed LMC calcite. Estimated seawater dolomite (ESD) compositions are reconstructed utilizing identical assumptions with respect to seawater and foraminifer calcite tests analyzed in Miller *et al.* (1987). K represents the ratio of seawater TDC to freshwater TDC. Models used equations in Banner and Hanson (1991). Seawater and freshwater fluid compositions used in these calculations are presented in Table 4.6.

at the 50% groundwater mixtures suggested by the $\delta^{18}O$ versus $\delta^{13}C$ modelling. Therefore, unlike the modern groundwaters of Curaçao, the Miocene groundwaters of Curaçao did not contain basalt-derived Sr. A possible explanation for this is that the weathered basement regolith profiles required to liberate Sr during water-rock interaction were not yet formed in the Middle Miocene due to insufficient time, rainfall, or other unknown reasons.

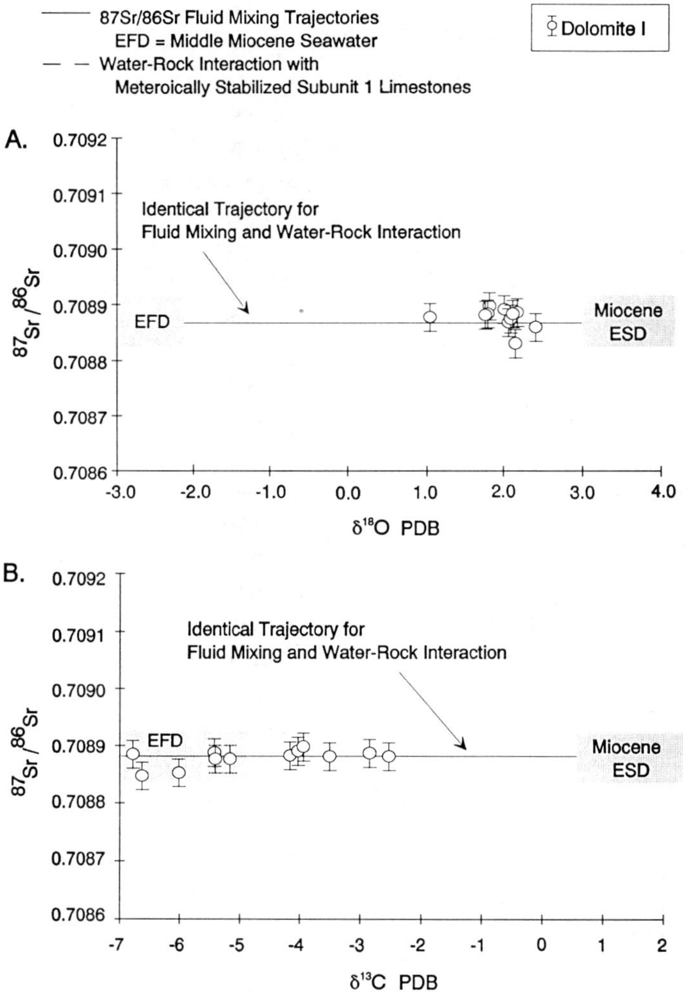

FIGURE 4.41. Evaluation of covariations in $^{87}Sr/^{86}Sr$ versus $\delta^{18}O$ and $\delta^{13}C$ compositions of Dolomite I. Seawater and freshwater fluid compositions used in these calculations are presented in Table 4.6. Mixing trajectories and water-rock interaction reaction pathways were modelled using equations presented in Banner and Hanson (1991). A. Crossplot of Dolomite I $\delta^{18}O$ versus $^{87}Sr/^{86}Sr$. Estimated freshwater dolomite (EFD) contains Sr with an $^{87}Sr/^{86}Sr$ composition equivalent to that of contemporaneous Middle Miocene seawater, therefore fluid mixing and water-rock interaction trajectories are identical. B. Crossplot of Dolomite I d18C versus $^{87}Sr/^{86}Sr$. EFD contains Sr with an $^{87}Sr/^{86}Sr$ composition equivalent to that of contemporaneous Middle Miocene seawater, therefore fluid mixing and water-rock interaction trajectories are identical.

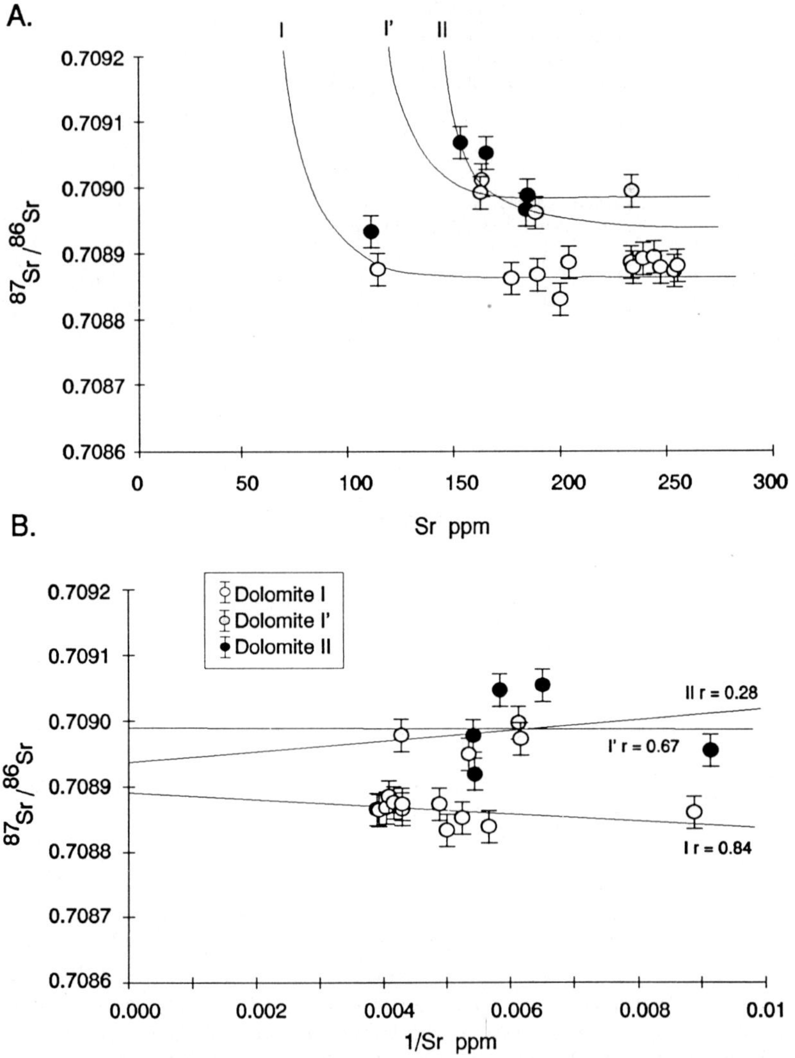

FIGURE 4.42. Covariations of $^{87}Sr/^{86}Sr$ versus Sr abundances in Dolomites I, I', and II. Data have been modelled using hyperbolic mixing equations presented in Langmuir *et al.* (1978). Mixing curves are calculated from the two most compositionally dissimilar samples of each type of dolomite. Close correlation of dolomite data on both the Sr and 1/Sr plots, and consistancy with other geochemical and geological evidence, can be used to indicate that the dolomite Sr was derived from the mixing of two different sources of Sr. A. Cross-plot of $^{87}Sr/^{86}Sr$ versus Sr. B. Cross-plot of $^{87}Sr/^{86}Sr$ versus 1/Sr.

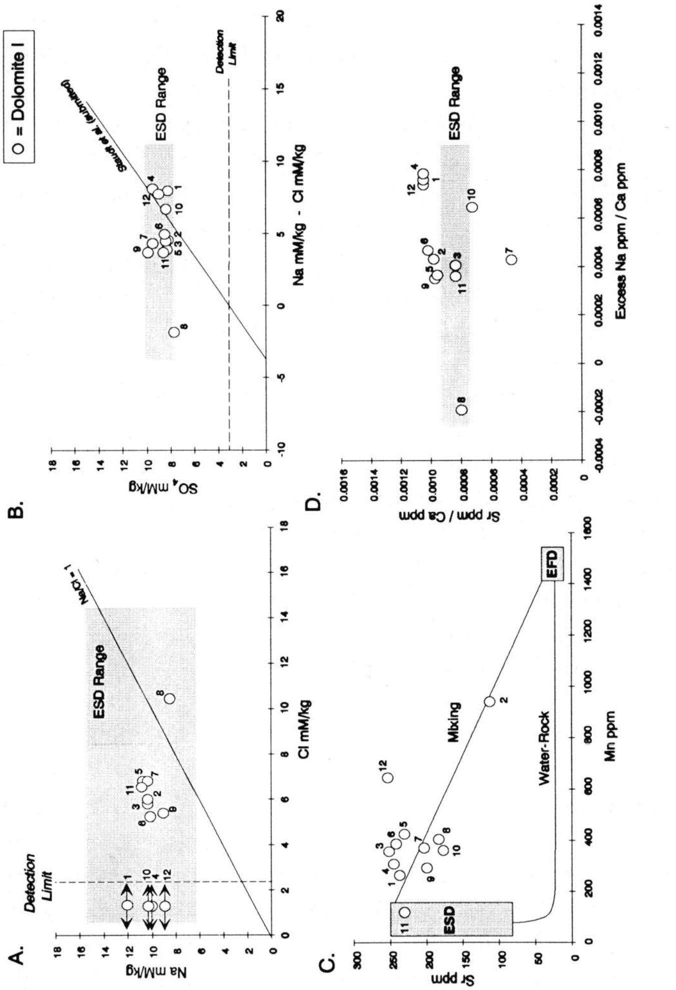

FIGURE 4.43. Evaluation of covariations in Dolomite I Na, SO_4, Cl, Sr, and Mn compositions. Sample numbers refer to individual dolomite analyses that are listed in Table 4.3. A. Crossplot of molar Cl versus molar Na compositions of. Molar Na/Cl line of 1 represents average compositions of seawater and mixed seawater-freshwater fluids. B. Crossplot of excess molar Na versus molar SO_4 compositions. Staudt *et al.* (1993) line represents trend observed in dolomites from other localities. C. Crossplot of Mn ppm versus Sr ppm compositions. Hypothetical mixing and water-rock trajectories between estimated seawater dolomite (ESD) and estimated freshwater dolomite (EFD) compositions calculated using equations presented by Banner and Hanson (1990). D. Crossplot of excess Na/ Ca ppm versus Sr/Ca ppm compositions.

Therefore the meteoric groundwaters did not contain the high Sr abundances necessary to alter the seawater-dominated $^{87}Sr/^{86}Sr$ signature of a seawater-freshwater mixture.

The Na, Cl and SO_4 compositions of Dolomite I are also consistent with precipitation from fluids derived from the mixing of seawater and meteoric groundwater. The low abundance of Na, Cl and SO_4 in Dolomite I with respect to concentrations in dolomites of hypersaline origin is consistent with precipitation in a seawater-freshwater mixing zone (Staudt *et al.*, 1993; Figs. 4.33 and 4.34). Dolomite I also contains molar Na/Cl ratios that are approximately equal to, or slightly greater than 1, which are similar to the Na/Cl of 1 in seawater and mixing zone fluids (Fig. 4.43). Seawater-freshwater mixing is further supported by the correlation of Dolomite I Mn versus Sr data with binary mixing lines between the ESD and EFD compositions rather than water-rock interaction trajectories (Fig. 4.43). Dolomite I data also exhibits moderately good correlation with mixing curves calculated for excess Na versus $\delta^{13}C$ and $^{87}Sr/^{86}Sr$ (Fig. 4.44).

Two dolomitization events have been observed in Dolomite I from CL-defined replacement fabrics within the dolomite-limestone contacts at Seru Blandan. These events include an initial cementation phase followed by simultaneous cementation and centripetal replacement (Fig. 4.16). This suggests dolomitizing fluid equilibria changed from saturation to under saturation with respect to calcite during the dolomitization process. Fluctuations in fluid equilibria of this type are expected in mixing zone dolomitization environments (e.g. Ward and Halley, 1985; Humphrey, 1988), with the magnitude and direction of shifts in saturation being dependent on the initial composition of the freshwater and the initial proportion of freshwater in the mixture (Badiozmani, 1973; Land, 1973a, 1985; Hardie, 1987; Stoessel *et al.*, 1989). Comparison with fluid equilibria observed in modern mixing environments of the Yucatan Peninsula (Stoessel *et al.*, 1989; Fig. 4.45) suggests that these changes from calcite saturation to undersaturation in Subunit 1 may represent either: 1) increasing proportions of seawater in freshwater-dominated fluids ($<20\%$ seawater); or 2) increasing proportions of freshwater in seawater-dominated fluids ($<20\%$ freshwater; Fig. 4.45). However, small errors in the estimated value for dolomite saturation could cause significant errors in these reconstructions, ranging from: 1) saturation at all degrees of mixing due primarily to the initial saturation of the freshwater with respect to dolomite; to 2) the onset of dolomite saturation at mixtures

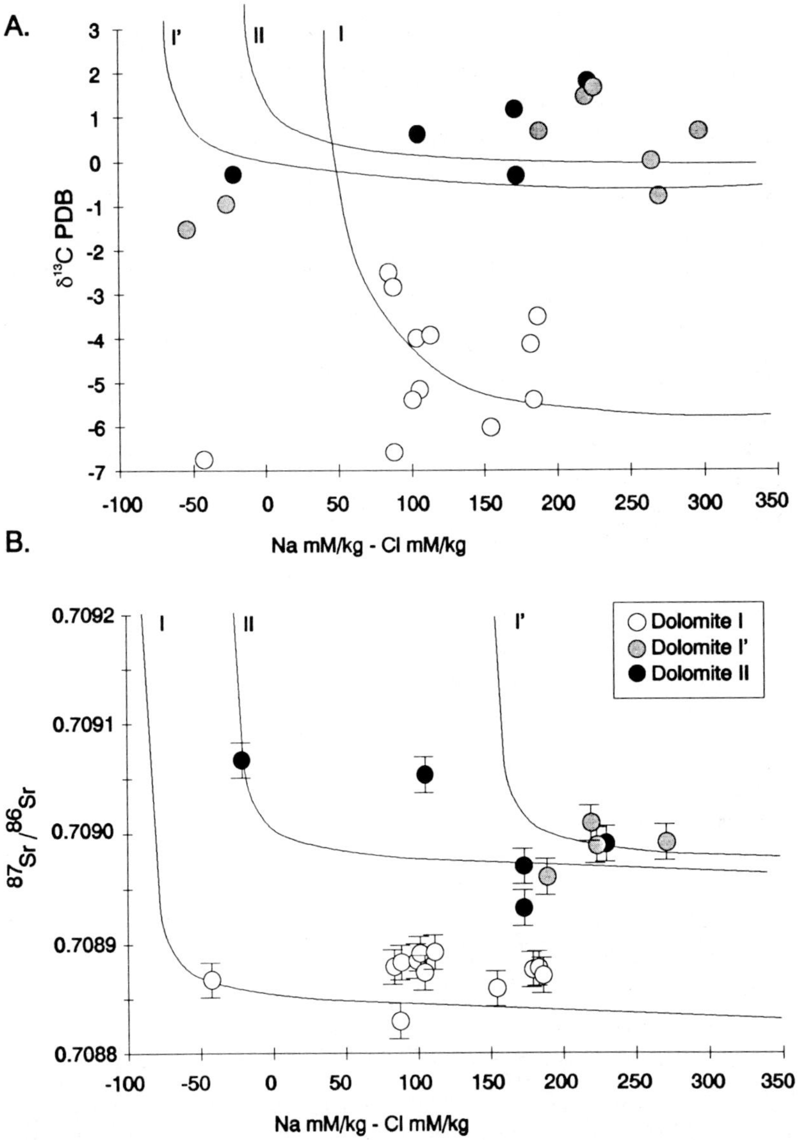

FIGURE 4.44. Evaluation of covariations in $^{87}Sr/^{86}Sr$, $\delta^{13}C$ and excess Na in Dolomite I, I', and II. Data have been modelled using hyperbolic mixing equations presented in Langmuir *et al.* (1978). Mixing curves are calculated from the two most compositionally dissimilar samples of each type of dolomite. A. Crossplot of excess Na in ppm versus $\delta^{13}C$ compositions. B. Crossplot of excess Na in ppm versus $^{87}Sr/^{86}Sr$ compositions.

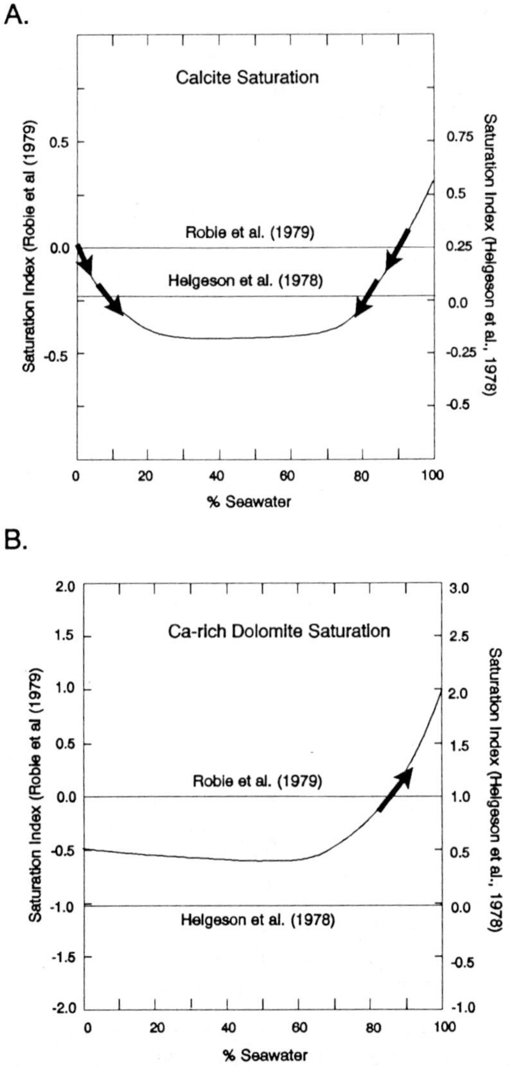

FIGURE 4.45. Saturation indices at 25°C and 1 atm for mixtures of meteoric groundwaters with seawater. Data and calculations from Stoessel *et al.* (1988) in the Yucatan Peninsula of Mexico. A and B exhibit mixing curves for calcite and Ca-rich dolomite saturation respectively, computed for increasing extents of seawater mixing with freshwater relative to the thermodynamic data of Helgeson *et al.* (1978) and Robie *et al.* (1979). Black arrows on each cross-plot illustrate relative extents of mixing associated with changes in groundwater carbonate equilibria.

of <20% freshwater (Fig. 4.45). Therefore, these uncertainties prevent a direct correlation of Dolomite I centripetal replacement with fluid saturation states during progressive mixing. However, the overall paragenetic sequence of pre-dolomite meteoric diagenesis followed by mixing zone dolomitization (Fig. 4.18) is suggestive of increasing proportions of seawater in the Subunit 1 aquifers with time.

Distribution of Dolomitization: The pervasive updip stratiform and bed cross-cutting distributions of Dolomite I exhibited at Seru Blandan and St. Michielsberg have generally been attributed to the gravity-driven hypersaline reflux of seawater-derived brines (Deffeyes *et al.*, 1965; Lucia and Major, in press). However, the above geochemical evaluations strongly suggest that Dolomite I was precipitated from reducing seawater-freshwater mixing zone fluids. Therefore, petrographic evidence from across the dolomite-limestone contacts will evaluate this apparent contradiction between the evidence for fluid composition and hydrology.

The overall tabular stratiform dolomite body (SDB) geometry mimics the bedding within Subunit 1 limestones, and indicates that precursor depositional bedding and facies distributions had a strong control on Dolomite I cement and replacement distributions (Figs. 4.13 and 4.14). "Halos" of pervasive and partial replacement dolomitization occur *above*, as well as *below* and *downdip* of each stratiform dolomite body, and both grainstones and wackestones are dolomitized (Fig. 4.15). These relationships suggest that while dolomite cementation was *dependent* upon precursor permeability, pervasive replacement dolomitization may have been relatively *independent* of permeability. Previous studies of dolomitization have assumed that "massive (pervasive replacement) dolomite records the flow path geometry of the fluid circulation system operational during dolomitization" (Wilson, 1989, p. 110). However, Dolomite I SDBs suggest that instead of exclusively representing conduits in which fluid flow rates were higher than within the surrounding lithologies, dolomite distributions more accurately represent regions in which higher extents of dolomitization reaction were achieved. The important distinction of this definition is that dolomitization reactions can be enhanced by a variety of parameters that may or may not have been a direct function of permeability and ensuing flow rates. Under these conditions, a given geometrically defined body of dolomite could represent either: 1) spatially selective enhancement of geochemical reactions within a lithologic section that was pervasively bathed in a single dolomitizing fluid; or 2) enhanced flow rates through a spatially distinct portion of a lithologic section, which may or may not have been pervasively saturated by a dolomitizing fluid. The distribution of Dolomite I precipitation from seawater-dominated mixing zone fluids appears to be a hybrid of these two processes. The overall updip distribution was controlled by the mixing zone interface between seawater and freshwater and its associated high flow rates (e.g. Simms, 1984), while the stratiform dolomite bodies are the result of local extents of replacement dolomitization reactions that may have been independent of enhanced flow rates.

Petrogenesis of Dolomite I'

Dolomite I' contains low abundances of Sr, high abundances of Mn and Fe, consistent $^{87}Sr/^{86}Sr$ signatures, and occurs in both Subunit 1 and 2. This evidence suggests that Dolomite I' was precipitated from seawater-dominated fluids at 4.8 to 5.2 ± 1.4 Ma during the late Late Miocene to Early Pliocene (Fig. 4.37). This Dolomite I' age is consistent with the paragenetic sequence of leaching, calcitization, and dolomitization in that it is significantly younger than the 13.3 ± 1.6 Ma age of Subunit 1 limestone stabilization, and is equivalent to or slightly younger than the biostratigraphically determined depositional age of Subunit 2 Lithofacies 5 (Fig. 4.37). The intraformational sealevel curve suggests that the Curaçao horst continued to be uplifted throughout the Miocene, elevating the Subunit 1 deposits during the Late Miocene to bathymetries consistent with the shallow water deposition of Subunit 2 Lithofacies 5 *Montastrea* boundstones (<100 m water depth; Fig. 4.38). These ages for pre-dolomite meteoric diagenesis and Dolomite I' precipitation correlate well with the third-order eustatic sea level lowstand and ensuing sea level rise during the Miocene-Pliocene transition (Fig. 4.38).

Composition of the Dolomitizing Fluids: The $\delta^{18}O$, $\delta^{13}C$ and $^{87}Sr/^{86}Sr$ compostions of Dolomite I' suggest that the dolomitizing fluids were composed of a mixture of seawater and meteoric groundwater. Dolomite I' $\delta^{18}O$ versus $\delta^{13}C$ data exhibit a good correlation with modelled trajectories representing precipitation from a mixture of seawater and freshwater rather than trajectories representing fluid alteration during water-rock interaction (Fig. 4.46). Al-

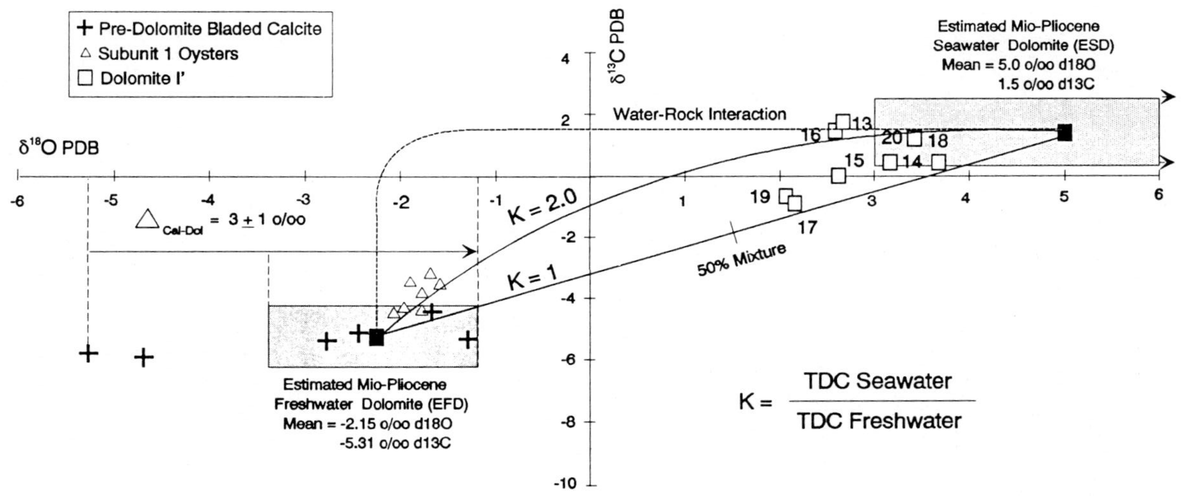

FIGURE 4.46. Crossplot of Dolomite I' $\delta^{18}O$ versus $\delta^{13}C$ compositions with respect to dolomite precipitation from mixed seawater and freshwater. Sample numbers are presented in Table 4.3. See Figure 4.40 for description.

though data points 13 and 16 contain slightly elevated $\delta^{13}C$ compositions, the other data exhibit covariations closely correlated with the mixing trajectories (Fig. 4.46). The intermediate position of the Dolomite I' $\delta^{18}O$ and $\delta^{13}C$ data between the ESD and EFD endmembers implies that these seawater-freshwater mixtures contained as much as 40% freshwater (Fig. 4.46). All Dolomite I' samples fall above the linear total dissolved carbon (TDC) mixing line between the ESD and EFD, which suggests that the seawater contained up to twice as much C as the freshwater with which it mixed ($K \leqslant 2.0$; Fig. 4.46).

Covariations in $^{87}Sr/^{86}Sr$ versus $\delta^{18}O$ and $\delta^{13}C$ were incapable of discriminating between fluid mixing and water-rock interaction due to the narrow range of Dolomite I' $^{87}Sr/^{86}Sr$ compositions (Fig. 4.47). However, no close correlations were observed between the Dolomite I' $^{87}Sr/^{86}Sr$ versus Sr and 1/Sr data and modelled mixing trajectories (Fig. 4.42). As was the case for Dolomite I, the radiogenic $^{87}Sr/^{86}Sr$ signatures of Dolomite I' suggest that the freshwaters that mixed with the seawater had low Sr abundances, and were therefore not directly analogous to groundwaters on modern Curaçao.

The Na, Cl, and SO_4 compositions of Dolomite I' are also consistent with precipitation from fluids derived from the mixing of seawater and meteoric groundwater. The low abundances of Na, Cl, and SO_4 in Dolomite I' with respect to dolomites precipitated from hypersaline brines is consistent with precipitation from a seawater-freshwater mixed fluid (Staudt *et al.*, 1993; Figs. 4.33 and 4.44). Dolomite I' contains molar Na/Cl ratios that are equal to, or slightly greater than the Na/Cl=1 of seawater (Fig. 4.43). And Dolomite I' also exhibits a good correlation with mixing curves calculated for excess Na versus $\delta^{13}C$ and $^{87}Sr/^{86}Sr$ (Fig. 4.43). Seawater-freshwater mixing is further supported by the correlation of Dolomite I Mn versus Sr data binary mixing lines between the ESD and EFD (Fig. 4.48).

As was the case for Dolomite I, two dolomitization events have been observed in Dolomite I' via CL-defined centripetal replacement fabrics across the dolomite-limestone contacts (Fig. 4.16). These petrographic relationships are consistent with fluctuations expected in mixing zone environments (i.e. Ward and Halley, 1985; Humphrey, 1988; Stoessel *et al.*, 1989). This is also consistent with increasing proportions of seawater the host Subunit 1 and 2 limestone aquifers with time, and correlates well with the relative rise in third-order sealevel at the Miocene-Pliocene transition (Figs. 4.37 and 4.38).

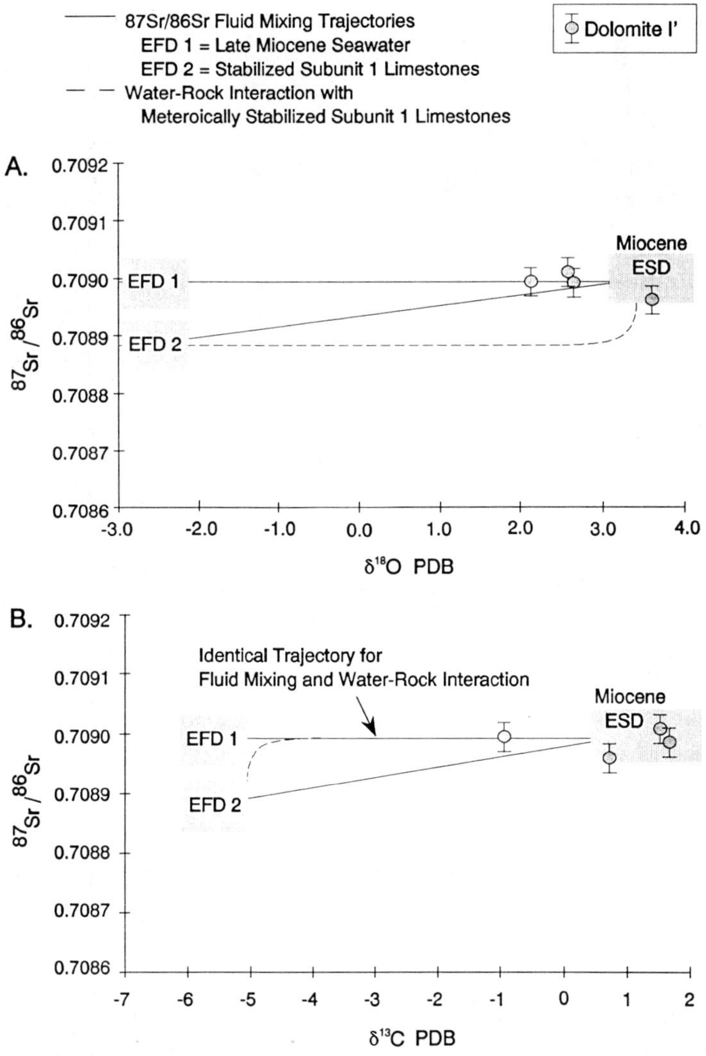

FIGURE 4.47. Evaluation of covariations in $^{87}Sr/^{86}Sr$ versus $\delta^{18}O$ and $\delta^{13}C$ compositions of Dolomite I'. Seawater and freshwater fluid compositions used in these calculations are presented in Table 4.6. Mixing trajectories and water-rock interaction reaction pathways were modelled using equations presented in Banner and Hanson (1991). A. Crossplot of $\delta^{18}O$ versus $^{87}Sr/^{86}Sr$. EFD 1 contains Sr with an $^{87}Sr/^{86}Sr$ composition equivalent to that of contemporaneous Late Miocene seawater. EFD 2 contains Sr with an $^{87}Sr/^{86}Sr$ composition equivalent to that of stabilized Subunit 1 limestones. B. Crossplot of d18C versus $^{87}Sr/^{86}Sr$.

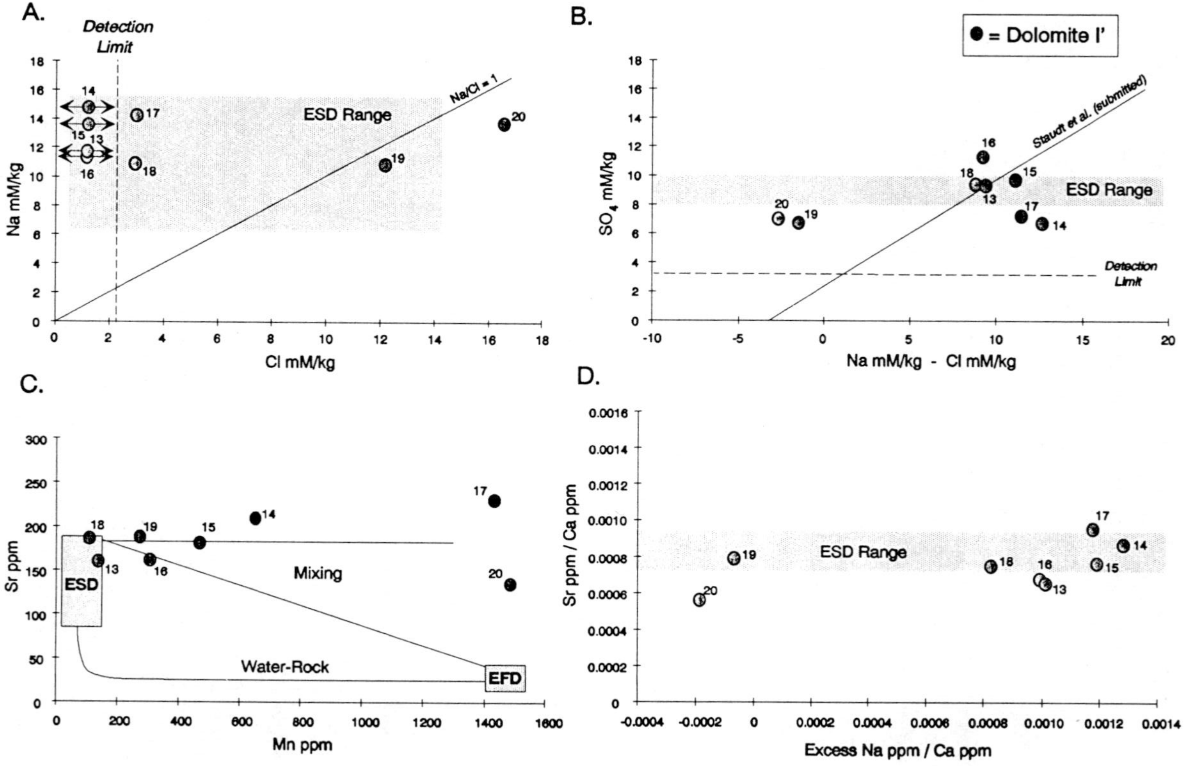

FIGURE 4.48. Evaluation of covariations in Dolomite I' Na, SO_4, Cl, Sr, and Mn compositions. See Fig. 4.43 for description.

Distribution of Dolomitization: The above geochemical evaluations suggest that Dolomite I' was precipitated from reducing seawater or seawater-dominated mixing zone fluids. The bed cross-cutting dolomitization exhibited by Dolomite I' is also in agreement with dolomitization models for mixing zone that have migrated through the lithology during changes in relative sealevel (e.g. Land, 1974) The only exception to pervasive bed cross-cutting Dolomite I' distributions is at Seru Prekstul. Here the lithologies retain some local areas of stratiform dolomitization, and are therefore consistent with the stratiform distributions observed for Dolomite I.

Pre-Dolomite II Diagenesis

Subunit 1 Lithofacies 7 limestones containing Dolomite II exhibit evidence for intense leaching of skeletal grains, replacement of oyster and brachiopod shells and precipitation of low-Mg calcite (LMC) blocky calcite cements. All of these features are suggestive of cementation and dissolution associated with meteoric and mixing zone diagenesis (e.g. Land, 1970; Matthews, 1974; James and Choquette, 1983; Budd, 1984; Saller and Moore, 1989). Continued uplift of the Curaçao horst brought the Seroe Domi Formation into water depths shallow enough by the Pliocene to permit the deposition of *in situ* Lithofacies 7 *Agaracia* boundstones (Fig. 4.38). Therefore, this tectonic-eustatic setting, in addition to the light O and C isotopes, enriched Fe and Mn, and depleted Sr compositions of Subunit 1 skeletal material, suggest that Subunit 7 diagenesis represents meteoric alteration rather than deep marine diagenesis (e.g. Saller and Koepnick, 1990).

Analyses of Subunit 2 Lithofacies 7 LMC oyster and brachiopod shells suggest that they were replaced by a LMC meteoric calcite with a $\delta^{18}O$ of approximately −0.75 to −2.25‰ PDB, which is equivalent to the initial meteoric calcite suggested by the Subunit 1 oyster material (Fig. 4.39). Therefore, this requires that the O isotopic composition of Curaçao groundwaters fluctuated significantly during the Mio-Pliocene, with shifts of −2.2 to −5.4 for Subunit 1 Lithofacies 1 limestones, and −6.9 to −2 for Subunit 2 Lithofacies 7 limestones (Fig. 4.39).

The modelling of the $^{87}Sr/^{86}Sr$ compositions in texturally unaltered, partially altered, and completely altered Subunit 2 oyster and brachiopod shells suggests that the LMC replacement calcite had Sr ratios of 0.70903 (Fouke, 1993). Although the Sr in these replacement calcites may have been derived from a variety of sources on Curaçao (Fig. 4.35), this implies a maximum age for Subunit 2 limestone stabilization of Pliocene to Early Pleistocene at approximately 1.95 $\pm$0.6 Ma (after Hodell *et al.*, 1991). This maximum age correlates with the third-order global sealevel draw downs during with the Plio-Pleistocene transition (Haq *et al.*, 1987; Figs. 4.37 and 4.38).

Petrogenesis of Dolomite II

The geochemical composition of Dolomite II is distinct from that of Dolomite I and I' in that: 1) Dolomite II $^{87}Sr/^{86}Sr$ values vary from being equivalent to less radiogenic than the host Subunit 2 limestones, exhibiting a spread in $^{87}Sr/^{86}Sr$ that is 2.75 times greater than the tightly clustered compositions of Dolomite I and I' (Fig. 4.36); and 2) the most nonradiogenic $^{87}Sr/^{86}Sr$ Dolomite II sample contains more than an order of magnitude more Zn and Cu than Dolomites I and I', while the most radiogenic Dolomite II sample contains an equivalently low abundance of Zn and Cu with respect to Dolomite I and I' (Table 4.3). This evidence suggests that the fluids precipitating Dolomite II varied in $^{87}Sr/^{86}Sr$, Zn, and Cu compositions. Of the four primary sources of Sr on Curaçao, only the weathered Cretaceous basalts of the Curaçao Lava Formation, and basalt-derived sands comprising the 250-m thick Subunit 2 Lithofacies 6 section, could have simultaneously provided less radiogenic Sr and high abundances of Zn and Cu. The possibility of sourcing Sr from the basement is supported by the less radiogenic $^{87}Sr/^{86}Sr$ values and high Sr abundances in modern groundwaters of the Curaçao Lava Formation (Tables 4.5 and 4.6). Therefore, several lines of evidence have been combined to suggest that Dolomite II precipitated from a mixture of seawater with a freshwater containing Sr derived from the basement, which include: 1) seawater-dominated trace element compositions; 2) a wide spread in $^{87}Sr/^{86}Sr$ that reaches values less radiogenic than the host Subunit 2 limestones; 3) the correlation of Zn and Cu enrichment with less radiogenic $^{87}Sr/^{86}Sr$ values; and 4) analogy with modern Curaçao groundwaters.

Age modelling of the most radiogenic Dolomite II $^{87}Sr/^{86}Sr$ value suggests precipitation from a Late Pliocene to Early Pleistocene seawater-dominated fluid at 1.9 $\pm$0.6 Ma (Fig. 4.36). The lack of Dolomite II within Subunit 3 and the Quaternary Limestone Terraces, and the multiple carbonate and basaltic sources of less radiogenic Sr on Curaçao during the Pliocene, combine to suggest that this is a maximum estimate of the age of precipitation.

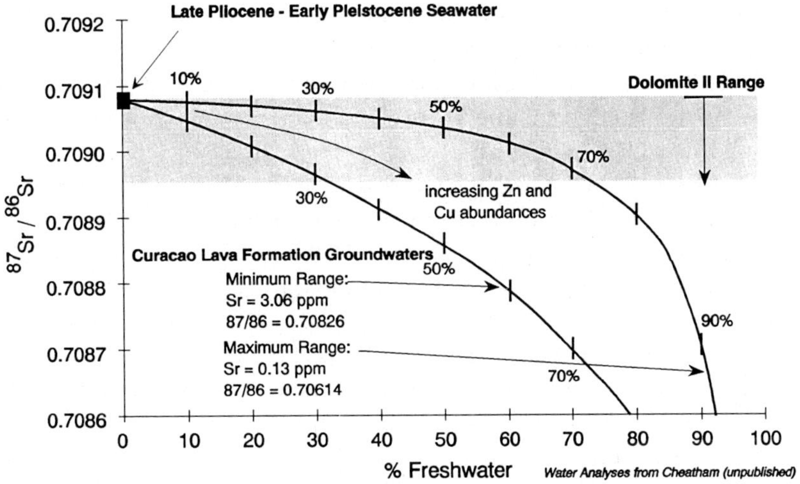

FIGURE 4.49. Variations in $^{87}Sr/^{86}Sr$ during the mixing of Plio-Pleistocene seawater with Curaçao Lava Formation groundwaters. Mixing trajectories were modelled using equations presented in Banner and Hanson (1991). Endmember fluid compositions are presented in Tables 4.5 and 4.6. The less radiogenic Sr compositions of Dolomite II could have been produced by as little as 30% freshwater mixed with seawater, due to the extremely high abundance of less radiogenic Sr in groundwaters of the Curaçao Lava Formation as measured by Cheatham (1984).

This interpreted age of Dolomite II precipitation is approximately equivalent to, the 1.9 to 2.2 ±0.6 Ma Late Pliocene estimate of Subunit 2 Lithofacies 7 limestone stabilization determined from altered brachiopod and oyster shells. These ages of Subunit 2 limestone diagenesis and Dolomite II precipitation correlate well with the third-order eustatic sealevel lowstand at the Pliocene-Pleistocene transition (Haq *et al.*, 1987; Figs. 4.37 and 4.38).

Composition of the Dolomitizing Fluids: Mass balance calculations of the mixing of Pleistocene seawater with modern Curaçao groundwaters suggests that mixtures of as little as approximately 30% freshwater would be necessary to produce dolomitizing fluids with $^{87}Sr/^{86}Sr$ values equivalent to those observed in Dolomite II (Fig. 4.49). In addition, groundwaters in other localities have mobilized significant amounts of Zn and Cu from weathered basalts (Freeze and Cherry, 1986; Aharon *et al.*, 1986). Therefore increasing amounts of Curaçao groundwaters would probably be accompanied by increasing enrichments of Zn, Cu, and less radiogenic Sr in the dolomitizing fluids. These systematic covariations are consistent with the ranges and correlation trends observed in Dolomite II.

The $\delta^{18}O$ and $\delta^{13}C$ compositions of Dolomite II further suggest that the dolomitizing fluids were composed of a mixture of seawater with meteoric groundwater. Dolomite I data correlate well with depletions in $\delta^{13}C$ and $\delta^{18}O$ predicted during the mixing of seawater and isotopically-depleted freshwater rather than with water-rock interaction trajectories (Fig. 4.50). The intermediate position of Dolomite II $\delta^{18}O$ and $\delta^{13}C$ data between the ESD and EFD endmembers implies that the seawater-freshwater fluid mixtures containing less than 50% freshwater (Fig. 4.50). Dolomite II sample analyses occur just below and above the linear TDC mixing line between the ESD and EFD, suggesting that seawater contained more C than the freshwater with which it mixed ($K \leqslant 2.0$; Fig. 4.50). Covariations in $^{87}Sr/^{86}Sr$ versus $\delta^{18}O$ and $\delta^{13}C$ correlate more closely with mixing lines between the ESD and a modern Curaçao groundwater EFD than with water-rock interaction pathways (Fig. 4.51). However, the Dolomite II $^{87}Sr/^{86}Sr$ versus Sr and 1/Sr covariations exhibit no consistent correlations (Fig. 4.42).

The Na, Cl, and SO_4 concentrations in Dolomite II are also suggestive of precipitation from fluids created by the mixing of seawater with meteoric groundwater. The low abundances of Na, Cl, and SO_4 with respect to dolomites precipitated in hypersaline environments is consistent with precipitation from mixed seawater and freshwater fluids (Staudt *et al.*, 1993; Figs. 4.33 and 4.34). Dolomite II also contains molar Na/Cl ratios that are equal to, or slightly greater than the Na/Cl=1 of seawater (Fig. 4.52). Dolomite II exhibits a good correlation with mixing curves calculated for excess Na versus $\delta^{13}C$ and $^{87}Sr/^{86}Sr$ (Fig. 4.43). Seawater-freshwater mixing is further supported by the correlation of Dolomite II Mn versus Sr data binary mixing lines between the ESD and EFD (Fig. 4.52).

Distribution of Dolomitization: The patchy distributions of Dolomite II, combined with the geochemical evidence summarized above, further suggests precipitation from a mixing zone circulation system. Randomly distributed updip pockets of dolomites containing small crystals are observed in Pleistocene mixing zones (e.g. Humphrey, 1988). These local distributions are an additional example of spatially selective areas of enhanced dolomitization reactions within a section that is pervasively bathed updip in dolomitizing waters. While similar distributions could be created by other hydrologic sys-

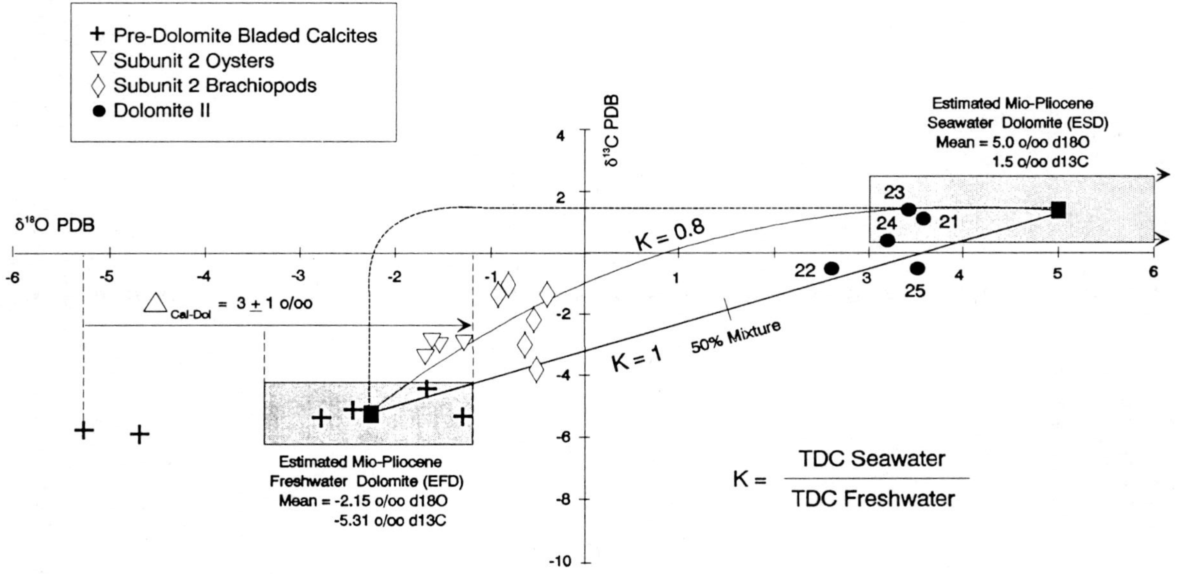

FIGURE 4.50. Modelling of Dolomite II $\delta^{18}O$ versus $\delta^{13}C$ compositions with respect to precipitation from mixed seawater-freshwater dolomitization fluids. Sample numbers refer to compositions presented in Table 4.3. See Figure 4.40 for description.

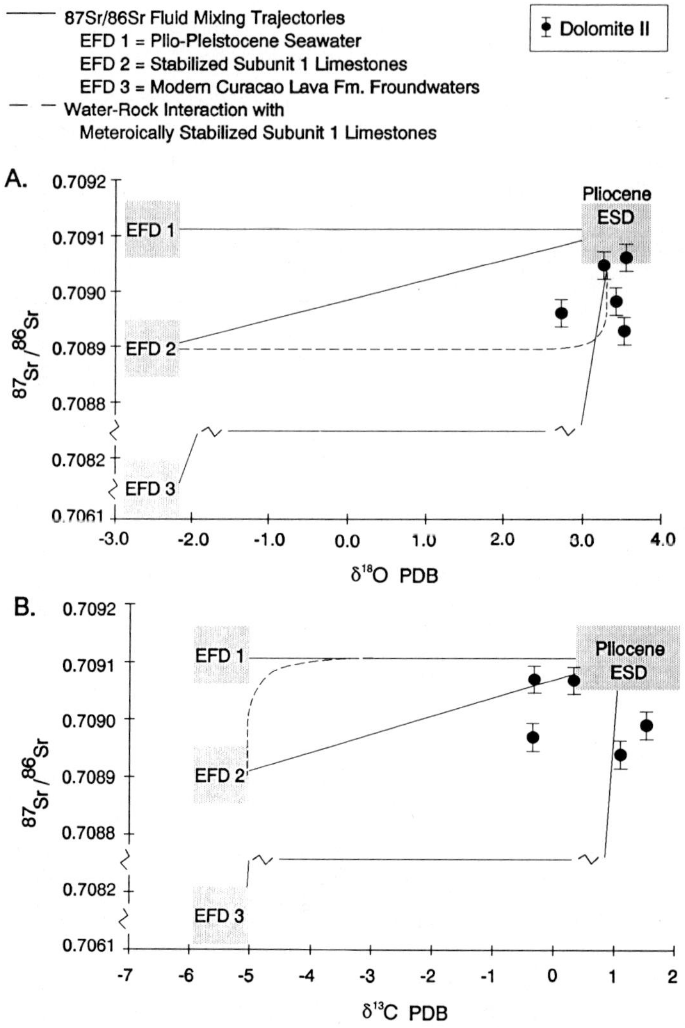

FIGURE 4.51. Evaluation of covariations in $^{87}Sr/^{86}Sr$ versus $\delta^{18}O$ and $\delta^{13}C$ compositions of Dolomite II. Seawater and freshwater fluid compositions used in these calculations are presented in Table 5.2. Mixing trajectories and water-rock interaction reaction pathways were modelled using equations presented in Banner and Hanson (1991). A. Crossplot of $\delta^{18}O$ versus $^{87}Sr/^{86}Sr$. Estimated freshwater dolomite (EFD) 1 contains Sr with an $^{87}Sr/^{86}Sr$ composition equivalent to that of contemporaneous Plio-Pleistocene seawater. EFD 2 contains Sr with an $^{87}Sr/^{86}Sr$ composition equivalent to that of stabilized Subunit 1 limestones. EFD 3 contains less radiogenic 87S6/86Sr equivalent to values observed in the modern Curaçao Lava Formation groundwaters. B. Crossplot of d18C versus $^{87}Sr/^{86}Sr$.

tems, the derivation of less radiogenic Sr, Zn, and Cu from basalts requires that the basalts be subaerially exposed and intensely weathered.

Post-Dolomite Diagenesis

The geochemistry of post-Dolomite I, I', and II blocky and columnar calcite cementation is consistent with diagenesis observed in meteoric and mixing zone environments (e.g. Land, 1973b; Matthews, 1974; James and Choquette, 1983; Budd, 1984; Saller and Moore, 1989). The significant depletions in isotopic ratios and trace element abundances (Fig. 4.21) are most consistent with changes in meteoric fluid chemistry in response to uptake of isotopically light soil-gas CO_2 and water-rock interaction (e.g. Meyers and Lohmann, 1985; Lohmann, 1987). These late stages of calcitization suggest that the $\delta^{18}O$ of calcite continued to become more depleted with respect to the two-stage meteoric stabilization processes that took place prior to dolomitization. Eventually the meteoric water values of approximately −6.0‰ PDB were reached (Fig. 4.21).

Conclusions

The Seroe Domi Formation contains three petrographically distinct types of dolomite, which have been designated as Dolomite I, I', and II. Dolomite I is composed of large crystals up to 300 μm in diameter that exhibit yellow-orange cathodoluminescent concentric crystal zoning. Dolomite I' is composed of similarly large crystals, yet they exhibit an orange-brown cathodoluminescent coarser-scale concentric crystal zoning as well as sectoral zoning. In contrast, Dolomite II is composed of significantly smaller crystals of up to 75 μm in diameter, which exhibit a dark CL core and an outer orange-brown bright cathodoluminescent rim.

Dolomite I, I', and II exhibit distinct stratigraphic and geographic distributions within the Seroe Domi Formation. Dolomite I occurs in Subunit 1 Lithofacies 1, 2, and 3 lithologies in the south-central portion of the island from Seru Kabritu to Seru Kura. Dolomite I' occurs in Subunit 1 Lithofacies 1, 3, and 4 lithologies, as well as in Subunit 2 Lithofacies 5 lithologies in the southeastern third of Curaçao from Seru Mansinga to Duivelsklip. And Dolomite II occurs in Subunit 2 Lithofacies 7 lithologies across most of the

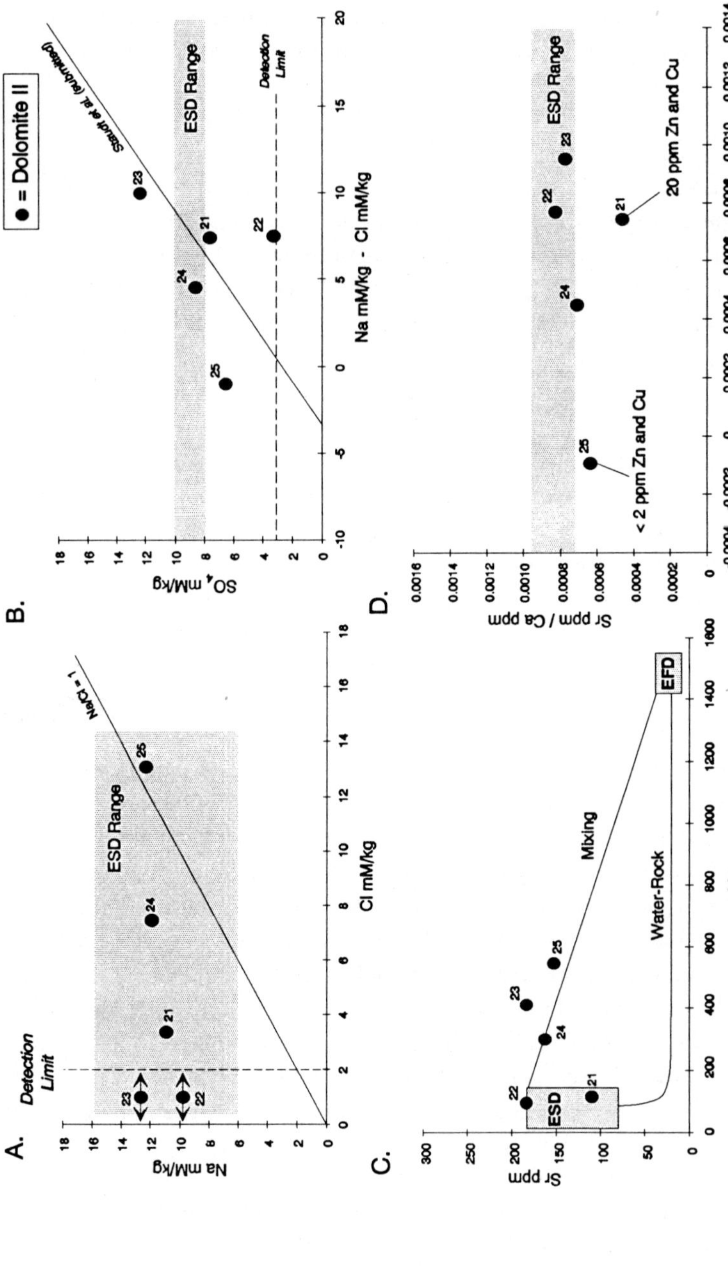

FIGURE 4.52. Evaluation of covariations in Dolomite II Na, SO_4, Cl, Sr, and Mn compositions. See Figure 4.43 for description.

length of Curaçao, from Seru Belefa to Kabrietenberg. Dolomite I occurs primarily as stratiform dolomite bodies (SDBs) within the lowermost 20 m of Subunit 1 sections, which mimic the strike and dip of the host limestones. Conversely, Dolomite I' occurs predominantly as bed cross-cutting dolomite bodies that have pervasively dolomitized the entire outcrop. And, unlike either Dolomite I or I', Dolomite II occurs as poorly developed localized pockets of pervasive dolomite.

The Seroe Domi Formation has experienced multiple episodes of limestone diagenesis and dolomitization that have successively taken place after each major depositional period. Petrographic relationships illustrate that meteoric leaching and low-Mg calcite bladed and blocky calcite cementation preceeded Dolomite I and I' precipitation. Similar evidence of extensive meteoric stabilization is observed in the Dolomite II host limestones. However no petrographic evidence of the relative timing of meteoric stabilization with respect to Dolomite II precipitation was observed. Post-dolomitization diagenesis has included dolomite dissolution as well as the precipitation of meteoric calcite, gypsum, and phosphate cements.

Petrographic analyses of the dolomite-limestone contacts at the margins of the Dolomite I stratiform dolomite bodies have tracked the evolution of the fluid-rock system by documenting the incipient stages of the dolomitization process. These analyses have shown that: 1) an initial phase of dolomite cementation predated a later episode of cementation and replacement; 2) in some places, the initial cements served as nucleation sites for replacement dolomitization; 3) the initial phase of dolomite cementation was strongly controlled by pre-existing porosity and permeability; 4) the later dolomitization phase, which was dominated by replacement, was relatively independent of pre-existing depositional and diagenetic facies distributions.

Modern groundwaters within the Curaçao Lava Formation exhibit significant enrichments in Ca, Mg, Cl and HCO_3, and simultaneous depletions in Na. In addition, analyses of similar Bonaire groundwaters exhibit significant Sr enrichments and extremely less radiogenic $^{87}Sr/^{86}Sr$ values compared with modern seawater. Evaluation of covariations in solute abundances in these anomalous groundwaters suggests that they may have been created by semi-closed system water-rock interaction of meteoric fluids with weathered basalts of the Curaçao Lava Formation.

Preservation of the internal consistency in orientation of large-scale with small-scale crystal growth features suggest that Dolomites I, I', and II are primary precipitates from solution. Therefore fluid chemistries can be directly interpreted from solid phase compositions. Dolomites I, I', and II contain consistently low abundances of Sr, Mn, Na, Cl, F, and SO_4. This suggests that Dolomites I, I', and II were precipitated from reducing fluids similar in composition to seawater or seawater-freshwater mixtures.

Covariations in Dolomite I $\delta^{18}O$, $\delta^{13}C$, $^{87}Sr/^{86}Sr$, and Sr compositions suggest that Dolomite I was precipitated from fluids created by the mixing of seawater with isotopically depleted freshwater. Dolomite I exhibits tightly clustered $^{87}Sr/^{86}Sr$ values that are more radiogenic than their host Middle Miocene Subunit 1 limestones, and consistent with precipitation during the late Middle Miocene to early Late Miocene at 9.4 to 111.3 ±1.4 Ma. This dolomitization age is slightly younger than the $^{87}Sr/^{86}Sr$ geochronology of Subunit 1 limestone stabilization, and is consistent with the paragenetic sequence. This sequence of Subunit 1 pre-Dolomite I meteoric diagenesis followed by Dolomite I precipitation correlates with the Late to Middle Miocene eustatic third-order sealevel lowstand associated with the onset of Antarctic glaciation, and the ensuing relative sealevel rise.

Dolomite I' compositional covariations in $\delta^{18}O$, $\delta^{13}C$, $^{87}Sr/^{86}Sr$, and Sr suggest that Dolomite I' was precipitated from fluids created by the mixing of seawater with isotopically-depleted freshwater. The tightly clustered Dolomite I $^{87}Sr/^{86}Sr$ values range from being equivalent to more radiogenic than their host Middle to Late Miocene limestones, and are consistent with precipitation during the Late Miocene to Early Pliocene at 4.8 to 5.2 ±1.4 Ma. This dolomitization age is significantly younger than the $^{87}Sr/^{86}Sr$ geochronology of Subunit 1 limestone stabilization, and probably slightly younger than the age of Subunit 1 Lithofacies 5 stabilization, and is therefore consistent with the paragenetic sequence. This sequence of Subunit 1 and 2 pre-Dolomite I' meteoric diagenesis followed by Dolomite I' precipitation correlates with a Late Miocene to Early Pliocene eustatic third-order sealevel lowstand and the ensuing relative sealevel rise.

Dolomite II compositional covariations in $\delta^{18}O$ and $\delta^{13}C$, differential enrichments in Zn and Cu, and range in $^{87}Sr/^{86}Sr$ suggest that Dolomite II was precipitated from fluids created by the mixing of seawater with isotopically-depleted freshwater. Unlike Dolomites I and I', Dolomite II exhibits a relatively wide variation in $^{87}Sr/^{86}Sr$ values that range from being *equivalent*, to significantly *less* radiogenic than the host Subunit 2 Lithofacies 7 limestones. In addition, the less radiogenic $^{87}Sr/^{86}Sr$ samples are significantly enriched in Zn and Cu, while the most radiogenic $^{87}Sr/^{86}Sr$ are depleted in these elements. Therefore, analogy with modern Curaçao Lava Formation groundwaters suggest that these Dolomite II Sr isotopic compositions represent a mixture between contemporaneous seawater Sr and a less radiogenic Sr derived from the Curaçao basement. Minimum estimates of the age of precipitation based on Sr isotope regression age modelling of the most radiogenic Dolomite II $^{87}Sr/^{86}Sr$ value, suggests precipitation during the Pliocene to Early Pleistocene at approximately 1.9 $\pm$0.6 Ma. This Dolomite II age is slightly younger than, to equivalent with, the $^{87}Sr/^{86}Sr$ geochronology of Subunit 1 limestone meteoric stabilization. In the absence of petrographic evidence, this suggests that dolomitization either immediately followed or was contemporaneous with limestone stabilization. This timing of Subunit 2 limestone diagenesis and Dolomite II precipitation correlates with a Late Pliocene to Early Pleistocene eustatic third-order sealevel lowstand and the ensuing relative sealevel rise.

REFERENCES

ABTMAIER, B. F., 1979. *Ground water investigation of Curaçao.* Unpublished report to the Department of Agriculture and Animal Husbandry, Island Government of Curaçao, Netherlands Antilles, 252 p.

ADEY, W. H. & BURKE, R. B., 1977. Holocene bioherms of Lesser Antilles- geologic control of development. *in* Frost, S. H. *et al.* (eds.), *Reefs and related carbonates – ecology and sedimentology.* Amer. Assoc. Petrol. Geologists Stud. Geol. 4: 67-81.

ADEY, W. H. & MACINTYRE, I.G., 1973. Crustose coralline algae; a re-evaluation in the geological sciences. *Bull. Geol. Soc. Amer.* 84: 883-903.

ADEY, W. H., TOWNSEND, R. A. & BOYKINS, W.T., 1982. The crustose coralline algae of the Hawaiian Archipelago. *Contr. Mar. Sci. Smith. Inst.* 15: 1-74.

AHARON, P., SOCKI, R. A. & CHAN, L., 1987. Dolomitization of atolls by seawater convection flow: Test of a hypothesis at Nieue, South Pacific. *Jour. Geol.* 95: 187-203.

AIZAWA, S. & AKAIWA, H. 1988. Geochemical behavior of transition metals during the formation of protodolomite in Minamidaitojima Island, Japan. *Chem. Geol.* 67: 275-284.

ALEXANDER, C.S., 1961. The marine terraces of Aruba, Bonaire, and Curaçao, Netherlands Antilles. *Assoc. Amer. Geograph. Ann.* 51: 102-123.

AUSTIN, J. A., *et al.*, 1985. *Bahamas.* Proceed. Ocean Drill. Prog. 101, 520 p.

BADIOZAMANI, K., 1973. The dorag dolomitization model – application to the middle Ordivician of Wisconsin. *Jour. Sed. Petrol.* 43: 965-984.

BAK, R. P. M., 1977, Coral reefs and their zonation in Netherlands Antilles. *in* Frost, S. H. *et al.* (eds.), *Reefs and related carbonates – ecology and sedimentology.* Amer. Assoc. Petrol. Geol. Stud. Geol. 4: 3-16.

BAK, R. P. M. & B. E. LUCKHURST, 1980. Constancy and change in coral reef habitats along depth gradients at Curaçao. *Oecologia* 47: 145-155.

BAKER, P.A. & KASTNER, M., 1981. Constraints on the formation of sedimentary dolomite. *Science* 213: 214-215.

BAKER, P.A. & BURNS, S.J., 1985. Occurence and formation of dolomite in organic-rich continental margin sediments. *Amer. Assoc. Petrol. Geol. Bull.* 69: 1917-1930.

BANDOIAN, C. A. & MURRAY, R.C., 1974. Pliocene-Pleistocene carbonate rocks of Bonaire, Netherlands Antilles. *Bull. Geol. Soc. Amer.* 85: 1243-1252.

BANNER, J. L. & HANSON, G. N., 1990. Calculation of simultaneous isotopic and trace element variations during water-rock interaction with applications to carbonate diagenesis. *Geochim. Cosmochim. Acta* 54: 3123-3127.

BANNER, J. L., HANSON, G. N. & MEYERS, W.J., 1988. Water-rock interaction history of regionally extensive dolomites of the Burlington-Keokuk Formation (Mississippian):

Isotopic evidence. *in* Shukla, V. & Baker, P.A. (eds.), *Sedimentology and Geochemistry of Dolostones*. Soc. Econ. Paleon. Mineral. Spec. Pub. 43: 97-114.

Barber, C., 1974. Major and trace element associations in limestones and dolomites. *Chem. Geol.* 14: 273-280.

Barnaby, R. & Rimstidt, J. D., 1989. Redox conditions of calcite cementation interpreted from Mn and Fe contents of authigenic calcites. *Bull. Geol. Soc. Amer.* 101: 795-804.

Beets, C.J., 1991. The late Neogene 87Sr/86Sr isotopic record in the western Arabian Sea, Site 722. *In* Prell, W.L., and N. Niitsuma (eds.), *Proceed. Ocean Drill. Prog. Sci. Res.* 117: 459-465.

Beets, C.J., 1992. *Calibration of late Cenozoic marine strontium isotope variations and its chronostratigraphic and geochemical applications.* Unpublished Ph.D. dissertation, Vrije Universiteit, Amsterdam, 133 p.

Beets, D.J. 1971. New data on the stratigraphy of Curaçao, Netherland Antilles. *in* Fifth Caribbean Geological Conference Transactions, *Queens Coll. Geol. Bull.* 5: 85-88.

Beets, D.J., 1972. *Lithology and stratigraphy of the Cretaceous and Danian succession of Curaçao.* Natuurw. Studiekring Sur. Ned. Ant., Utrecht, 70, 153 p.

Beets, D.J., 1977. Cretaceous and Early Tertiary of Curaçao. *Field Guide 8th Carib. Geol. Conf. (Curaçao)* 15: 7-17.

Beets, D.J., Maresch, W.V., Klaver, G.T., Monttana, A., R. Bocchio, Beunk, F.F. & Monen, H.P., 1984. Magmatic rock series and high pressure metamorphism as constraints on the tectonic history of the Southern Caribbean, *in* W.E. Bonini, R.B. Hargraves and R. Shagam, (eds.), *The Caribbean-South American Plate Boundary and Regional Tectonics*, Geol. Soc. Amer. Mem. 162: 95-130.

Berggren, W.A., Kent, D.V., Flynn, J.J. & Van Couvering, J.A., 1985. Cenozoic geochronology. *Bull. Geol. Soc. Amer.* 96: 1407-1418.

Biju-Duval, B., Mascle, A., Rosales, H. & Young, G., 1982. Episutural Oligo-Miocene basins along the Northern Venezuela margin, *in* Watkins, J.S., and Drake, C.L., (eds.), *Studies in Continental Margin Geology*, Amer. Assoc. Petrol. Geol. Mem. 34:347-358.

Bosellini, A. & Ginsburg, R.N., 1971. Form and internal structure of recent algal modules (rhodolites), Bermuda. *Jour. Geol.* 79: 669-682.

Bosence, D.W.J., 1983. Coralline algal reef frameworks. *Journ. Geol. Soc. London* 140: 365-376.

Bosence, D.W.J., 1984. Construction and preservation of two recent algal reefs, St. Croix, Caribbean. *Palaeontology* 27: 549-574.

Bosence, D.W.J., 1985. Preservation of coralline-algal reef frameworks. *in Proc. 5th Intern. Coral Reef Cong.* (Tahiti) 6: 623-628.

Bosence, D.W.J., 1991. Corraline Algae: mineralization, taxonomy, and palaeoecology. *In* Riding, R., (ed.), *Calcareous Algal and Stromatolites*, Springer-Verlag, p. 98-113.

Bosence, D.W.J. & Pedley, H.M., 1982. Sedimentology and palaeoecology of a Miocene coralline algal biostome from the Maltese Islands. *Palaeogeography* 38: 9-43.

Budd, A.F., Stemann, T.A. & Johnson, K.G., in press. Stratigraphic distributions of genera and species of Neogene to Recent Caribbean reef corals. *Journ. Paleonotology* 68.

Budd, D.A., 1984. *Freshwater diagenesis of Holocene ooid sands, Schooner Cays, Bahamas.* Unpublished Ph.D. dissertation, University of Texas at Austin, Texas, 491 p.

Burke, W.H., Denison, R.E., Hetherington, E.A., Koepnik, R.B., Nelson, H.F. & Otto, J.B., 1982. Variation of seawater 87Sr/86Sr throughout Phanerozoic time. *Geology* 10: 516-519.

BUSENBERG, E. & PLUMMER, L.N., 1985. Kinetic and thermodynamic factors controlling the distribution of SO^{2-} and Na^{+} in calcites and selected aragonites. *Geochim. Cosmochim. Acta* 49: 713-725.

CADASTRAL SURVEY DEPARTMENT OF THE NETHERLANDS ANTILLES, 1982. *Topographic Maps of Curaçao, Netherlands Antilles.* Scale of 1:25,000.

CANDER, H.S., KAUFMAN, J., DANIELS, L.D., MEYERS, W.J., 1988. Regional dolomitization of shelf carbonates in the Burlington-Keokuk formation (Mississippian), Illinois and Missouri: Constraints from cathodoluminescent zonal stratigraphy. *In* Shukla, V. and Baker, P.A., (eds.), *Sedimentology and Geochemistry of Dolostones*, Soc. Econ. Paleon. Mineral. Spec. Pub. 43: 129-144.

CARBALLO, J.D., LAND, L.S. & MISER, D.E., 1987. Holocene dolomitization of supratidal sediments by active tidal pumping, Sugarloaf Key, Florida. *Jour. Sed. Petrol.* 57: 153-165.

CASE, J.E., 1974. Geophysical Studies in the Caribbean Sea. *in* A.E.W. Nairn and F.G. Stahli, (eds.), *The Ocean Basins and Margins,* 3, Plenum Press, New York. p. 107-173.

CASE, J.E. & HOLCOMBE, T.L., 1980. Geologic-tectonic map of the Caribbean region. *United States Geological Survey Miscellaneous Investigations* Map I-1100. Scale of 1:2,500,000.

CASE, J.E., HOLCOMBE, T.L. & MARTIN, R.G., 1984. Map of geological provinces in the Caribbean region. *in* W.E. Bonini, R.B. Hargraves and R. Shagam, (eds.), *The Caribbean-South American Plate Boundary and Regional Tectonics*, Geol. Soc. Amer Mem. 162: 1-30.

CHEATHAM, M.M., 1984. *Sr isotope geochemisty of ocean water, Solar Salt Lake and Bonaire.* Unpublished manuscript, University of New Hampshire, New Hampshire, 25 p.

CHOQUETTE, P.W. & STEINEN, R.P., 1980. Mississippian non-supratidal dolomite, Ste. Genevieve Limestone, Illinois Basin: Evidence for mixed water dolomitization. *In* D.H. Zenger, *et al.*, (eds.), *Concepts and Models of Dolomitization*, Soc. Econ. Paleon. Mineral. Spec. Pub. 28: 163-196.

COOK, H.E., 1983. Ancient carbonate platform margins, slopes, and basins. *Society of Economic Paleontologists and Mineralogists*, Short Course Notes 12: 5.1-5.189.

CREVELLO, P.D. & SCHLAGER, W., 1980. Carbonate debris sheets and turbidiies, Exuma Sound, Bahamas. *Jour. Sed. Petrol.* 50: 1121-1148.

DE BUISONJÉ, P.H., 1974. *Neogene and Quaternary Geology of Aruba, Curaçao and Bonaire.* Natuurw. Studiekring Sur. Ned. Ant., Utrecht, 78, 293 p.

DE BUISONJÉ, P.H. & ZONNEVELD, J.I.S., 1976. Caracasbaai: A submarine slide of a huge coastal fragment in Curaçao. *Nieuwe West-Indische Gids* 5: 55-88.

DEFFEYES, K.S., LUCIA, F.T. & WEYL, P.K., 1965. Dolomitization of Recent and Plio-Pleistocene sediments by marine evaporite waters on Bonaire, Netherlands Antilles. *In* L.C. Pray and R.C. Murry, (eds.), *Dolomitization and Limestone Diagenesis*, Soc. Econ. Paleon. Mineral. Spec. Pub. 13: 71-80.

DEPAOLO, D.J., 1986. Detailed record of the Neogene Sr isotopic evolution of seawater from DSDP Site 59OB. *Geology* 14: 103-106.

DEPAOLO, D.J. & FINGER, K.L., 1991. High-resolution strontium-isotope stratigraphy and biostratigraphy of the Miocene Monterey Formation, central California. *Bull. Geol. Soc. Amer.* 103: 112-124.

DEPAOLO, D.J. & B.L. INGRAM, 1985. High-resolution with strontium isotopes. *Science* 227: 938-941.

DREVER, J.I., 1982. *The Geochemistry of Natural Waters.* Prentice-Hall, 388 p.

DICKSON, J.A.D., SMALLEY, P.C., RAHEIM, A., STIJFHOORN, D.E., 1990. Intracrystalline carbon and oxygen isotope variations in calcite revealed by laser microsampling. *Geology* 18: 809-811.

DICKSON, J.A.D., SMALLEY, P.C. & KIRKLAND, B.L., 1991. Carbon and oxygen isotopes in Pennsylvanian biogenic and abiogenic aragonite (Otero County, New Mexico): A laser microprobe study. *Geochim. Cosmochim. Acta* 55: 2607-2613.

EKDALE, A.A., BROMLEY, R.G., PEMBERTON, S.G., 1984. *Ichnology. The use of trace fossils in sedimentology and stratigraphy.* Soc. Econ. Paleon. Mineral. Short Course 15, 317 p.

FAURE, G. & POWELL, J.L., 1972. *Strontium Isotope Geology.* Springer-Verlag, 188 p.

FAURE, G., 1986. *Principles of Isotope Geology.* John Wiley and Sons, New York. 589 p.

FOCKE, J.W., 1978a. Holocene development of coral fringing reefs leeward off Curaçao and Bonaire (Netherlands Antilles). *Marine Geology* 28: 31-41.

FOCKE, J.W., 1978b. Subsea (0-40m) terraces and benches, windward off Curaçao, Netherland Antilles. *Leidse Geol. Meded.* 51: 95-102.

FOLK, R.L., 1965. Some aspects of recrystallization in ancient limestones. *in* Pray, L.C. and Murray, R.C., (eds.), *Dolomitization and limestone diagenesis: A symposium.* Soc. Econ. Paleon. Mineral. Spec. Pub. 13: 14-48.

FOLK, R.L. & LAND, L.S.. 1975. Mg/Ca ratio and salinity: Two controls over crystallization of dolomite. *Amer. Assoc. Petrol. Geol. Bull.* 59: 60-68.

FOUKE, B.W., 1993. *Chronostratigraphy and Dolomitization of the Seroe Domi Formation, Curaçao, Netherlands Antilles.* Unpublished Ph.D. dissertation, State University of New York at Stony Brook, New York, 850 p.

FOUKE, B.W. & LABARBARA, M.C., 1986. *Ecology of shallow-water brachiopods in Jamaica: Proof of predation and its implication* (abst.). 4th N. Amer. Paleon. Conf., Boulder, Colorado. p. A16.

FOUKE, B.W. & REEDER, R.J., 1992. Surface structural controls on dolomite composition: Evidence from compositional sectoral zoning. *Geochim. Cosmochim. Acta* 56: 4015-4024.

FRITZ, P. & KATZ, A., 1972. The sodium distribution of dolomite crystals. *Chem. Geol.* 10: 237-244.

FROST, S.H., 1977. Miocene to Holocene evolution of Caribbean Province reef-building corals: *In Proceed. Third Inter. Coral Reef Symp.*, p. 354-359.

FULTHORPE, C.S. & MELILLO, A.J., 1985. Middle Miocene carbonate gravity flows in the Straits of Florida at Site 626. *in* Austin, J.A., *et al.*, (eds.), *Proceed. Ocean Drill. Prog., Bahamas*, 101: 179-192.

GOLDSMITH, J.R. & GRAF, D.L., 1958. Relation between lattice constraints and composition of Ca-Mg carbonates. *Amer. Mineral.* 43: 84-101.

GRAF, C.H., 1972. Sedimentos del Terciario Superior u Cuaternario de la peninsula de Macanao, Margarita, Venezuela. *Trans. Sixth Carib. Geol. Conf.* (Caracas), Venezuela, p. 414-417.

GRAMMAR, G.M. & GINSBURG, R.N., 1992. Highstand versus lowstand deposition on carbonate platform margins: insight from Quaternary foreslopes in the Bahamas. *Mar. Geol.* 103: 125-136.

GUNATILAKA, A., AL-ZAMEL, A., SHEARMAN, D.J. & REDA, A., 1987. A spherulitic fabric in selectively dolomitized siliciclastic crustacean burrows, Northern Kuwait. *Jour. Sed. Petrol.* 57: 922-927.

HALLEY, R.B. & LUDWIG, K.R., 1989. Comment and reply on "Strontium-isotope stratigraphy of Enewetak Atoll" – Reply. *Geology* 17: 192-193.

HAQ, B.U., HARDENBOL, J. & VAIL, P.R., 1987. Chronology of fluctuating sealevels since the Triassic. *Science* 235: 1156-1166.

HARDIE, L.A. 1987. Dolomitization: a critical review of some current views. *Jour. Sed. Petrol.* 57: 166-183.

HEIN, J.R., GRAY, S.C., RICHMOND, B.M. & WHITE, L.D., 1992. Dolomitization of Quaternary reef limestone, Aitutaki, Cook Islands. *Sedimentology* 39: 645-661.

HEMMING, N.G., MEYERS, W.J. & GRAMS, J.C., 1989. Cathodoluminesence in diagenetic calcites: the roles of Fe and Mn as deduced from electron probe and spectrophotometric measurements. *Jour. Sed. Petrol.* 59: 404-411.

HERWEIJER, J.P. & J.W. FOCKE, 1878. Late Pleistocene depositional and denudational histroy of Aruba, Bonaire, Curaçao (Netherland Antilles). *Geol. Mijnbouw* 57: 177-187.

HESS, H. & MAXWELL, J.C., 1953. Caribbean research project. *Bull. Geol. Soc. Amer.* 64: 1-6.

HESS, J., BENDER, M.L. & SCHILLING, J.G., 1986. Evolution of the ratio of Srontium-87 to Strontium-86 in seawater from Cretaceous to present. *Science* 23: 979-983.

HINE, A.C., 1983. *Modern shallow water carbonate platform margins.* Soc. Econ. Paleon. Mineral. Short Course 12: 3.1-3.100.

HODELL, D.A., MEAD, G.A. & MUELLER, P.A., 1990. Variation in the strontium isotopic composition of seawater (8 ma to present); Implications for chemical weathering rates and dissolved fluxes to the oceans. *Chem. Geol.* 80: 291-307.

HODELL, D.A., MUELLER, P.A. & GARRIDO, J.R., 1991. Variation in the strontium composition of seawater during the Neogene. *Geology* 19: 24-27.

HODELL, D.A., MUELLER, P.A., MCKENZIE, J.A. & MEAD, G.A., 1989. Strontium isotope stratigraphy and geochemistry of the late Neogene ocean. *Ear. Planet. Sci. Let.* 92: 165-178.

HUBBARD, D.K., 1986. Sedimentation as a control of reef development: St. Croix, U.S.V.I. *Coral Reefs* 5: 117-125.

HUBBARD, D.K., MILLER, A.I. & SCATURO, D., 1990. Production and cycling of Calcium Carbonate in a shelf-edge reef system (St. Croix, U.S. Virgin Islands): Applications to the nature of reef systems in the fossil record. *Jour. Sed. Petrol.* 60: 335-360.

HUBBARD, D.K., BURKE, R.P. & GILL, I.P., 1985. Accretion in deep shelf-edge reefs, St Croix, U.S.V.I. *In* Crevello, P.D. & Harris, P.M., (eds.), *Deep Water Carbonates*, Soc. Econ. Paleon. Mineral. Core Workshop 6: 491-527.

HUBBARD, D.K., BURKE, R.P. & GILL, I.P., 1986. Styles of reef accretion along a steep, shelf-edge reef, St. Croix, U.S. Virgin Islands. *Jour. Sed. Petrol.* 56: 848-861.

HUBBARD, D.K., SADD, J.L. & ROBERTS, H.H., 1981. The role of physical processes in controlling sediment transport patterns on the insular shelf of St. Croix, U.S. Virgin Islands. *Proceed. Fourth Intern. Coral Reef Symp.* 1: 399-404.

HUBBARD, D.K., SUCHNEK, T.H., GILL I.P., COWPER, S., OGDEN, J.C., WESTERFIELD, J.R. & BAYES, J., 1981. *Proceed. Fourth Intern. Coral Reef Symp.* 1: 383-387.

HUMPHREY, J.D., 1988. Late Pleistocene mixing zone dolomitization, southeastern Barbados, West Indies. *Sedimentology* 35: 327-348.

HUMPHREY, J.D., 1989. Reply: Late Pleistocene mixing zone dolomitization, southeastern Barbados, West Indies. *Sedimentology* 36: 1138-1142.

ISHIKAWA, M. & ICHIKUNI, M., 1984. Uptake of sodium and potassium by calcite. *Chem. Geol.* 42: 137-146.

JACKSON, J.B.C., GOREAU, T.F. & HARTMAN, W.D., 1971. Recent brachiopod-coralline sponge communities and their paleoecological significance. *Science* 173: 623-625.

JACOBSON, R.L. & USDOWSKI, H.E., 1976. Partitioning of strontium between calcite, dolomite, and liquids. *Contrib. Mineral. Petrol.* 59: 171-185.

JAMES, N.P. & CHOQUETTE, P.W., 1983. Diagenesis 5. Limestones – The meteoric diagenetic environment. *Geosci. Canada* 10: 161-194.

JAMES, N.P. & CHOQUETTE, P.W., 1986. Diagenesis 6. Limestones – The sea floor diagenetic environment. *Geosci. Canada* 10: 162-178.

JAMES, N.P., GINSBURG R.N., MARSZALEK, D.S. & CHOQUETTE, P.W., 1976. Facies and fabric specificity of early subsea cements in shallow Belize (British Honduras) reefs. *Jour. Sed. Petrol.* 46: 523-544.

JAMES, N.P. & GINSBURG, R.N., 1979. The seeward margin of Belize barrier and atoll reefs. *Intern. Assoc. Sediment.* Spec. Pub. 3, 191 p.

JAMES, N.P., 1979. Reefs. *in* R.G. Walker, (ed.), *Facies Models.* Geosci. Canada Repr. Ser. 1: 121-133.

KATZ, A., 1973. The interaction of magnesium with calcite during crystal growth at 25°-90°C and one atmosphere. *Geochim. Cosmochim. Acta* 37: 1563-1586.

KATZ, A. & MATTHEWS, A., 1977. The dolomitization of $CaCO_3$: An experimental study at 252-295°C. *Geochim. Cosmochim. Acta* 41: 279-308.

KELLOGG, J.N. & BONINI, W.E., 1982. Subduction of the Caribbean Plate and basement uplift in the overriding South American Plate. *Tectonics* 1: 251-276.

KENTER, J.A.M., 1990. Carbonate platform flanks: slope angle and sediment fabric. *Sedimentology* 37: 777-794.

KINSMAN, D.J.J., 1969. Interpretation of Sr^{+2} concentrations in carbonate minerals and rocks. *Jour. Sed. Petrol.* 39: 486-508.

KITANO, Y., OKUMURA, M. & IDOGAKI, M., 1980. Abnormal behaviors of copper (II) and zinc ions in parent solution at the early stage of calcite formation. *Geochim. Cosmochim. Acta* 14: 167-175.

KITANO, Y., TOKUYAMA, A. & KANAMORI, N., 1968. Measurement of the distribution coefficient of zinc and copper between carbonate precipitate and solution. *Jour. Earth Sci. Nagoya Univ.* 16: 1-102.

KLAVER, G.T., 1987. *The Curaçao Lava Formation: An ophiolitic analogue of the anomalous thick layer 2B of the Mid-Cretaceous oceanic plateaus in the Western Pacific and Central Caribbean.* GUA Papers of Geology, Utrecht, Series 1, 27, 168 p.

KOEPNICK, R.B., BURKE, W.H., DENISON, R.E., HETHERINGTON, E.A., NELSON, H.F., OTTO, J.B. & WAITE, L.E., 1985. Construction of the seawater $^{87}Sr/^{86}Sr$ curve for the Cenozoic and Cretaceous: Supporting data. *Chem. Geol.* 58: 55-81.

KOPP, O.C., REEVES, D.K., RIVERS, M.L. & SMITH, J.V., 1990. Synchrotron X-ray fluoresence analysis of zoned carbonate gangue in Mississippi Valley-type deposits. *Chem. Geol.* 81: 337-347.

LAGAAY, R.A., 1969. Geophysical investigations of the Netherlands Antilles. *Ver. Kon. Ned. Acad. Wetensch. Afd. Natuurk., 1ste Reeks,* 5 (2), 86 p.

LAND, L.S., 1973a. Contemporaneous dolomitization of Middle Pleistocene reefs by meteoric water, North Jamaica. *Bull. Mar. Sci.* 23: 64-92.

LAND, L.S., 1973b. Holocene meteoric diagenesis of Pleistocene limestones, North Jamaica. *Sedimentology* 20: 411-424.

LAND, L.S., 1979. The fate of reef-derived sediment on the north Jamaican slope. *Marine Geology* 29: 55-71.

LAND, L.S., 1980. The isotopic and trace element geochemistry of dolomite: The state of the art. *In* Zenger, D.H. *et al.* (eds.). *Concepts and Models of Dolomitization,* Soc. Econ. Paleon. Mineral. Spec. Pub. 28: 87-110.

LAND, L.S., 1983. The application of stable isotopes to studies of the origin of dolomite and to problems of diagenesis of clastic sediments. *in* M.A. Arthur, *et al.*, (eds.), *Stable Isotopes in Sedimentary Geology,* Soc. Econ. Paleon. Mineral. Short Course Notes 10: 4.1-4.22.

LAND, L.S. 1985. The origin of massive dolomite. *Jour. Geol. Edu.* 33: 112-125.

LAND, L.S. & EPSTEIN. S., 1970. Late Pleistocene diagenesis and dolomitization, North Jamaica. *Sedimentology* 14: 187-200.

LAND, L.S. & HOOPS, G.K., 1973. Sodium in carbonate sediments and rocks: a possible index to the salinity of diagenetic solutions. *Jour. Sed. Petrol.* 43: 614-617.

LAND, L.S. & MOORE, C.H. JR., 1977. Deep forereef and upper island slope, north Jamaica. *in* S.H. Frost, Weiss, M.P., and Sauders J.B., (eds.), *Reefs and Related Carbonates-Ecology and Sedimentology.* Amer. Assoc. Petrol. Geol. Stud. Geol. 4: 53-65.

LAND, L.S. & MOORE, C.H. JR., 1980. Lithification, micritization, and syndepositional diagenesis of biolithites on the Jamaican island slope. *Jour. Sed. Petrol.* 50: 357- 369.

LANGMUIR, C.H., VOCKE, R.D. & HANSON, G.N., 1978. A general mixing equation with applications to Icelandic basalts. *Ear. Planet. Sci. Let.* 37: 380-392.

LOHMANN, K.C. 1987. Geochemical patterns of meteoric diagenetic systems and their application to studies of paleokarst. *in* Choquette, P.W. and James, N.P., (eds.), *Paleokarst*, Springer-Verlag, 415 p.

LU, F-Q., SMITH, J.V., SUTTON, S.R., RIVERS, M.L. & DAVIS, A.M., 1989. Synchrotron X-ray fluoresence analysis of rock-forming minerals. *Chem. Geol.* 75: 123-143.

LUCIA, F.J. & MAJOR, R.P., in press. Porosity occlusion by hypersaline reflux dolomitization. *Intern. Assoc. Sediment. Spec. Pub.*

LUCIA, F.J. & MAJOR, R.P., 1990. Porosity destruction by near-surface of Plio-Pleistocene carbonates, Bonaire, Netherland Antilles (abst.) *in Abstr. 13th Intern. Sediment. Cong.*, Nottingham, England.

LUCIA, F.J., 1968. Recent sediments and diagenesis of South Bonaire, Netherlands Antilles. *Jour. Sed. Petrol.* 38: 845-858.

LUDWIG, K.R., HALLEY, R.B., SIMMONS, K.R. & PETERMAN, Z.E., 1988. Strontium-isotope stratigraphy of Enewetak Atoll. *Geology* 16: 173-177.

LUMSDEN, D.N. & CHIMAHUSKY, J.S.,. 1980. Relationship between dolomite non-stoichiometry and carbonate facies parameters. *in* Zenger, D.H., *et al.*, (eds.), *Concepts and Models in Dolomitization*, Soc. Econ. Paleon. Mineral. Spec. Pub. 28: 123-138.

LUMSDEN, D.N. & LLOYD, R.V., 1988. An update of ESR spectroscopy in studies of dolomite origin. *In* Shukla, V., and Baker, P.A., (eds.), *Sedimentology and Geochemistry of Dolostones*, Soc. Econ. Paleon. Mineral. Spec. Pub. 43: 3-10.

LUMSDEN, D.N., SHIPE, L.G. & LLOYD, R.V., 1989. Mineralogy and Mn geochemistry of laboratory synthesized dolomite. *Geochim. Cosmochim. Acta* 53: 2325-2329.

MACHEL, H.G. & ANDERSON, J.H., 1989. Pervasive subsurface dolomitization of the Nisku Formation in central Alberta. *Jour. Sed. Petrol.* 59: 891-911.

MACHEL, H.G. & BURTON, E.A., 1991. Factors governing cathodoluminesence in calcite and dolomite, and their implications for studies of carbonate diagenesis. *In* Barker, C.E. and Kopp O.C. (eds.), *Luminesence Microscopy: Quantitative and Qualitative Aspects*, Soc. Econ. Paleon. Mineral. Short Course, p. 37-57.

MACINTYRE, I.G., 1967. Recent sediments off the west coast of Barbados, W.I. Unpublished P.h.D. dissertation, McGill University. 169 p.

MACINTYRE, I.G., 1988. Modern coral reefs of Western Atlantic: A new geological perspective. *Amer. Assoc. Petrol. Geol. Bull.* 72: 1360-1369.

MAJOR, R.P., LLOYD, R.M. & LUCIA, F.J., 1992. Oxygen isotope composition of Holocene dolomite formed in a humid hypersaline setting. *Geology* 20: 586-588.

MANN, P., C. SCHUBERT & K. BURKE, 1990. Review of Caribbean neotectonics. *in* Dengo, G., and Case, J.E., (eds.), *The Caribbean Region*, Geol. Soc. Amer. DNAG, H: 307-338.

MARESCH, W.V. 1974. Plate tectonics origin of the Caribbean Mountain System of Northern South American: Discussion and Proposal. *Bull. Geol. Soc. Amer.* 85: 669-682.

MATTHEWS, R.K., 1974. A process approach to diagenesis of reefs and associated limestones. *In* LaPorte, L. F. (ed), *Reefs in Time and Space*, Soc. Econ. Paleon. Mineral. Spec. Pub. 18: 234-256.

McIlreath, I. A. & James, N. P., 1978. Faces Models 12: Carbonate slopes. *Geosci. Canada* 5: 189-199.

Melillo, A.J., 1985. Neogene Planktonic Foraminifer biostratigraphy, ODP leg 101, Bahamas. *In* Austin, J.A., *et al.*, (eds.), *Proceed. Ocean Drill. Prog. Bahamas*, 101: 3-46.

Meyers, W.J., 1974. Carbonate cement stratigraphy of the Lake Valley Formation (Mississippian), Sacramento Mountains, New Mexico. *Jour. Sed. Petrol.* 44: 837-861.

Meyers, W.J., 1991. Calcite cement stratigraphy: An overview. *in* Barker, C. E. and Kopp, O.C., (eds.), *Luminesence Microscopy: Quantitative and Qualitative Apects*, Soc. Econ. Paleon. Mineral. Short Course, p. 133-147.

Meyers, W.J. & Lohmann, K.C., 1985. Isotope geochemistry of regiuonally extensive calcite cement zones and marine components in Mississippian limestones. *in* Schneiderman, N. & Harris, P.M., (eds.), *Carbonate Cements*, Soc. Econ. Paleon. Mineral. Spec. Pub. 36: 223-239.

Miller, K.G., Fairbanks, R.G. & Mountain, G.S., 1987. Tertiary oxygen isotope synthesis, sea level history, and continental margin erosion. *Paleooceanography* 2: 1-19.

Miller, K.G., Feigenson, M.D., Kent, D.V. & Olsson, R.K., 1988. Upper Eocene to Oligocene isotope (87Sr/86Sr, $\delta^{18}O$, $\delta^{13}C$) Standard Section, Deep Sea Drilling Project Site 522. *Paleooceanography* 3: 223-233.

Miller, K.G., Feigenson, M.D., Wright, J.D. & Clement, B.M., 1991. Miocene isotope reference section, Deep Sea Drilling Project Site 608: An evaluation of isotope and biostratigraphic resolution. *Paleooceanography* 6: 33-52.

Morelock, J., Grove, K. & Herandez, M.L., 1983. Oceanography and patterns of shelf sediments Mayaguez, Puerto Rico. *Jour. Sed. Petrol.* 53: 371-381.

Morse, J.W. & Bender, M.L., 1990. Partition coefficients in calcite: Examination of factors unfluencing the validity of experimental results and their application to natural systems. *Chem. Geol.* 82: 265-277.

Muessig, K.W., 1984. Structure and Cenozoic tectonics of the Falcon Basin, Venezuela, and adjacent areas. *in* Bonini, W.E., Hargraves, R.B., and Shagam, R., (eds.), *The Caribbean-South American Plate Boundary and Regional Tectonics*, Geol. Soc. Amer. Mem. 162: 217-230.

Mullins, H.T., 1979. Deep carbonate bank margin structure and sedimentation in the Northern Bahamas. *Society of Economic Paleontologists and Mineralogists*, Special Publication 27: 165-192.

Mullins, H.T., 1983. Modern carboate slopes and basins of the Bahamas. *Society of Economic Paleontologists and Mineralogists*, Short Course Notes 12: 4.1-4.138.

Mullins, H.T., 1985. Modern deep-water carbonates along the Blake-Bahama boundary. *in* Crevello, P.D., and Harris, P.M., (eds.), *Deep water Carbonates*, Soc. Econ. Paleon. Mineral. Core Workshop 6: 461-490.

Murray, R.C., 1969. Hydrology of south Bonaire, Netherlands Antilles – A rock selective dolomitization model. *Jour. Sed. Petrol.* 39:1007-1013.

Okumura, M. & Kitano, Y., 1986. Coprecipitation of alkali metal ions with calcium carbonate. *Geochim. Cosmochim. Acta* 50: 49-58.

O'Neil, J.R. & Epstein, S., 1966. Oxygen isotope fractionation in the system dolomite-calcite-carbon dioxide. *Science* 152: 198-201.

Oswald, E.J., 1992. *Dolomitization of a Miocene Reef Complex, Mallorca, Spain.* Unpublished Ph.D. dissertation, State University of New York at Stony Brook, New York, 350 p.

Paquette, J., 1992. Synchrotron X-ray fluoresence microanalysis of synthetic calcite crystals. *in* Hulbert, S.L., and Lazaraz, N., (eds.), *National Synchrotron Light Source Annual Report 1991*, Vol. 2, p. 300.

PEREZ, O.J. & AGGARVAL, Y.P., 1991. Present-day tectonics of the southeastern Caribbean and northeastern Venezuela. *Jour. Geophys. Res.* 86: 1753-1770.

PFLUM, C.E. & FRERICHS, W.E., 1976. Gulf of Mexico deep-water foraminifers. *Cushman Foundation for Foraminiferal Research*, Special Publication 14, 75 p.

PIERSON, B.J., 1981. The control of cathodoluminesence in dolomite by iron and manganese. *Sedimentology* 28: 601-610.

PINDELL, J. & DEWEY, J.F., 1982. Permo-Triassic reconstruction of western Pangea and the evaluation of the Gulf of Mexico-Caribbean region. *Tectonics* 1: 179-211.

PINGITORE, N.E. 1978. The behavior of Zn and Mn during carbonate diagenesis: theory and applications. *Jour. Sed. Petrol.* 48: 799-814.

PINGITORE, N.E., LYTLE, F.W., DAVIES, B.M., EASTMAN, M.P., ELLER, P.G. & LARSON, E.M., 1992. Mode of incorporation of Sr^{2+} in calcite: Determination by X-ray absorption spectroscopy. *Geochim. Cosmochim. Acta* 56: 1531-1538.

PLUMMER, L.N., 1975. Mixing of seawater with calcium carbonate groundwater. *In* Whitten, E.H.T., (ed.), *Quantiative Studies in Geological Sciences*, Geol. Soc. Amer. Mem. 142: 219-236.

PROSPERO, J.M., BONATTI, E., SCHUBERT, C. & CARLSON, T.N., 1970. Dust in the Caribbean atmosphere traced to the Africa dust storm. *Ear. Planet. Sci. Let.* 9: 287-293.

REEDER, R.L. & PROSKY, J.L., 1986. Compositional sector zoning in dolomite. *Jour. Sed. Petrol.* 56: 237-247.

RICHTER, F.M. & DEPAOLO, D.J., 1987. Numerical models for diagenesis and the Neogene Sr isotope evolution of seawater from DSDP Site 590B. *Ear. Planet. Sci. Let.* 83: 27-38.

ROEDDER, E., 1984. *Fluid Inclusions.* Mineral. Soc. Amer. Rev. Mineral. 12, 384 p.

ROWBOTTOM, R.J., 1978. *Report of Well Surveys on Curaçao.* Unpublished report to the Department of Agriculture and Animal Husbandry, Island Government of Curaçao, Netherlands Antilles.

ROWBOTTOM, R.J. & WINKEL, C.W., 1979. *Well Survey, Bonaire: Supplement I.* Unpublished report to the Department of Agriculture and Animal Husbandry, Island Government of Curaçao, Netherlands Antilles.

RUDE, P.D. & ALLER, R.C., 1991. Fluorine mobility during early diagenesis of carbonate sediment. An indicator of mineral transformations. *Geochim. Cosmochim. Acta* 55: 2491-2509.

RUNDBERG, Y. & SMALLEY, P.C., 1989. High-resolution dating of Cenozoic Sediments from Northern North Sea using 87Sr/86Sr stratigraphy. *Amer. Assoc. Petrol. Geol. Bull.* 73: 298-308.

SALLER, A.H., 1984. Petrologic and geochemical constraints on the origin of subsurface dolomite, Enewetok Atoll: An example of dolomitization by normal seawater. *Geology* 12: 217-220.

SALLER, A.H. & MOORE, C.H., JR., 1989. Meteoric diagenesis, marine diagenesis, and microporosity in Pleistocene and oligocene limestones, Enewetok Atoll, Marshall Islands. *Sed. Geol.* 63: 253-272.

SALLER, A.H. & KOEPNICK, R.B., 1990. Eocene to early Miocene growth of Enewetak Atoll: Insights from strontium isotope data. *Bull. Geol. Soc. Amer.* 102: 381-390.

SANTAMARIA, F. & SCHUBERT, C., 1974. Geochemistry and chronology of the Southern Caribbean- Northern Venezuela plate boundary. *Bull. Geol. Soc. Amer.* 7: 1085-1098.

SASS, E. & BEIN, A., 1989. Dolomites and salinity: A comparative geochemical study.

In Shukla, V., and Baker, P.A., (eds.), *Sedimentology and Geochemistry of Dolostones.* Soc. Econ. Paleon. Mineral. Spec. Pub. 43: 223-234.

SCHAUB, H.P., 1948. Geological observations on Curaçao, N.W.I. *Amer. Assoc. Petrol. Geol. Bull.* 32: 1275-1291.

SCHLAGER, W. & CHERMAK, A., 1979. Sediment facies of platform-basin transition, Tongue of the Ocean, Bahamas. *Soc. Econ. Paleon. Mineral. Spec. Pub.* 27: 193-208.

SCHUBERT, C., VALASTRO, S., COWARD, J.B., 1977. Evidencias de levantamiento reciente de la costa morte-central (Cordillera de la Costa), Venezuela. *Acta Cientos Venezuela.* 28: 363-372.

SCHUBERT, C. & VALASTRO, S., 1976. Quaternary geology of La Orchila Island, central Venezuelan offshore, Caribbean Sea. *Bull. Geol. Soc. Amer.* 87: 1131-1142.

SCHUBERT, C. & SZABO, B.J., 1978. Uranium-Series ages of Pleistocene marine deposits on the islands of Curaçao, and La Blanquilla, Caribbean Sea. *Geol. Mijnbouw* 57: 325-332.

SEILACHER, A., 1978. Use of trace fossil assemblages for recognizing depostional environments. *Soc. Econ. Paleon. Mineral. Short Course* 5: 167-81.

SHACKLETON, N.J. & KENNETT, J.P., 1975. Late Cenezoic oxygen and carbon isotopic changes at DSDP Site 284: Implications for the glacial history of the Northern Hemisphere and Antarctica. *in* Kennett, J., *et al.*, (eds.), *Init. Rep. Deep Sea Drill. Proj.* 29: 801-807.

SHACKLETON, N.J. & CITA, M.B., 1979. Oxygen and carbon isotope stratigraphy of benthic foraminifers at Site 397: Detailed history of climatic change during the Late Neogene. *in* Kennett, J., *et al.*, (eds.), *Init. Rep. Deep Sea Drill. Proj.* 29: 433-445.

SIBLEY, D.F., 1980. Climatic control of dolomitization, Seroe Domi Formation (Pliocene), Bonaire, Netherland Antilles. *in* D.H. Zenger, J.B. Dunham and R.L. Ethington, (eds.), *Concepts and Models of Dolomitization*, Soc. Econ. Paleon. Mineral. Spec. Pub. 28: 247-258.

SIBLEY, D.F., 1982. The origin of common dolomite fabrics: Clues from the Pliocene. *Jour. Sed. Petrol.* 52: 1087-1100.

SILVER, E.A., CASE, J.E. & MACGILLAVRY, H.J., 1975. Geophysical study of the Venezuelan borderland. *Bull. Geol. Soc. Amer.* 86: 213-226.

SIMMS, M., 1984. Dolomitization by groundwater flow systems in carbonate platforms. *Gulf Coast Assoc. Geol. Soc. Trans.* 34: 411-420.

SMALLEY, P.C., STIJFHORN, D.E., RAHEIM, A., JOHANSEN, H., & DICKSON, J.A.D., 1989. The laser microprobe and its application to the study of C and O isotopes in calcite and aragonite. *Sed. Geol.* 9: 257-273.

SOMMER, S.E., 1972. Cathodoluminesence in carbonates, 1. Characterization of cathodoluminesence from carbonate solid solutions. *Chem. Geol.* 9: 257-273.

STAUDT, W., OSWALD, E.J. & SCHOONEN, M.A.A., 1993. Determination of sodium, chloride and sulfate in dolomites: A new technique to constrain the composition of dolomitizing fluids. *Chem. Geol.* 107: 97-109.

STIENSTRA, P., 1983. Structure and general chemistry of Table Mountain Santa Barbara phosphates, Curaçao, Netherland Antilles. *Geol. Mijnbouw* 62: 275-284.

STIENSTRA, P., 1991. *Sedimentary petrology, origin and mining history of the phosphate rocks of Klein Curaçao, Curaçao, and Aruba, Netherlands West Indies.* Natuurw. Studiekring Sur. Ned. Ant., Utrecht, 130. 207 p.

STOESSEL, R.K., 1992. Effects of sulfate reduction on $CaCO_3$ dissolution and precipitation in mixing-zone fluids. *Jour. Sed. Petrol.* 62: 873-880.

STOESSELL, R.K., WARD, W.C., FORD, B.H. & SCHUFFERT, J.D., 1989. Water chemistry and

$CaCO_3$ dissolution in the saline part of an open-flow mixing zone, coastal Yucatan Peninsula, Mexico. *Bull. Geol. Soc. Amer.* 101: 159-169.

STUMM, W. & MORGAN, J.J., 1981. *Aquatic Chemistry.* John Wiley and Sons, 780 p.

SWART, P.K., RUIZ, J. & HOLMES, C.W., 1987. Use of strontium isotopes to constrain the timing and mode of dolomitization of Upper Cenozoic sediments in a core from San Salvador, Bahamas. *Geology* 15: 262-265.

SWART. P.K., BURNS, S.J. & LEDER, J.J., 1991. Fractionation of the stable isotopes of oxygen and carbon dioxide during the reaction of calcite with phosphoric acid as a function of temperature and technique. *Chem. Geol.* 86: 89-96.

TAKANO, B., 1985. Geochemical implications of sulfate in sedimentary carbonates. *Chem. Geol.* 5: 393-403.

TAKANO, B., ASANO, Y. & WATANUKI, K., 1980. Characterization of sulfate in travertine. *Contrib. Mineral. Petrol.* 72:197-203.

TEN HAVE, T., HEIJNEN, W. & NICKEL, E., 1982. Alterations in guano phospates and Mio-Pliocene carbonates of Table Mountain Santa Barbara, Curaçao. *Sed. Geol.* 31: 141-165.

TSUESUE, A. & HOLLAND, H.D., 1966. The coprecipitation of cations with $CaCO_3$-III. *Geochim. Cosmochim. Acta* 30: 439-453.

VAHRENKAMP, V.C., 1988. *Constraints on the formation of platform dolomites: A geochemical study of Late Tertiary dolomite from Little Bahama Bank, Bahamas.* Unpublished Ph.D. dissertation, Rosentiel School of Marine and Atmospheric Sciences, University of Miami, Florida, 328 p.

VAHRENKAMP, V.C., SWART, P.K. & RUIZ, J., 1988. Constraints and interpretation of $^{87}Sr/^{86}Sr$ ratios in Cenozoic dolomites. *Geophys. Res. Let.* 15: 385-388.

VAHRENKAMP, V.C. & SWART, P.K., 1990. New distribution coefficient for the incorporation of strontium into dolomite and its implications for the formation of ancient dolomites. *Geology* 18: 387-391.

VAN DUYL, F.C., 1985. *Atlas of the living reefs of Curaçao and Bonaire, Netherlands Antilles.* Natuurw. Studiekring Sur. Ned. Ant.,Utrecht, 117, 267 p.

VEIZER, J., 1983. Trace elements and isotopes in sedimentary carbonates. *in* Reeder, R.J., (ed.), *Carbonates: Mineralogy and Chemistry,* Mineral. Soc. Amer. Rev. Mineral. 11: 265-300.

VEIZER, J., 1983. Chemical diagenesis of carbonates: theory and application of trace element technique. *in* M.A. Arthur, *et al.*, (eds.), Stable Isotopes in Sedimentary Geology: *Soc. Econ. Paleon. Mineral. Short Course Notes* 10: 3-100.

VEIZER, J., LEMIEUX, B., JONES, B., GIBLING, M.R. & SAVELLE, J., 1977. Sodium: Paleosalinity indicator in ancient carbonate rocks. *Geology* 5: 177-179.

VIGNALI, M., 1972. Excursion geolgica al extremo occidental de Araya. *Trans. Sixth Carib. Geol. Conf.*, Caracas, Venezuela: 44-47.

WARD, W.B., 1993. *Marine Diagenesis of a Devonian Reef Complex, Canning Basin, Western Australia.* Unpublished Ph.D. dissertation, State University of New York Stony Brook, New York, 510 p.

WARD, W.B. & REEDER, R.J., 1993. The use of growth microfabrics and transmission electron microscopy in understanding replacement processes in carbonates. *in* Rezak, R., and Lavoie, D., (eds.), *Carbonate Microfabrics,* Springer-Verlag: 253-264.

WARD, W.C. & HALLEY, R.B., 1985. Dolomitization in a mixing zone of near seawater composition, late Pleistocene, N.E. Yucatan Peninsula. *Jour. Sed. Petrol.* 55: 407-420.

WEBER, J.N., 1964. Trace element composition of dolostones and dolomites and its bearing on the dolomite problem. *Geochim. Cosmochim. Acta* 28: 1817-1868.

WILSON, E.N., 1989. *Dolomitization of the Triassic Latemar buildup, Northern Italy.* Unpublished

Ph. D. dissertation, John Hopkins University, Maryland, 272 p.

WILSON, E. N., HARDIE, L. A. & PHILLIPS, O. M., 1990. Dolomitization front geometry, fluid flow patterns, and the origin of massive dolomite: The Triassic Latemar buildup, Northern Italy. *Amer. Jour. Sci.* 290: 741-796.

WOGELIUS, R. A., FRAZER, D. J., FELTHAM, D. J. & WHITEMAN, M. I., 1992. Trace element zoning in dolomite: proton microprobe data and thermodynamic constraints on fluid compositions. *Geochim. Cosmochim. Acta* 56: 319-334.

UITGAVEN NATUURWETENSCHAPPELIJKE STUDIEKRING VOOR HET CARAÏBISCH GEBIED

PUBLICATIONS OF THE FOUNDATION FOR SCIENTIFIC RESEARCH IN THE CARIBBEAN REGION

Obtainable from Secretariat Found. Sci. Res., Inst. Tax. Zool., Plantage Middenlaan 45, 1018 DC Amsterdam, The Netherlands, tel. & fax: 020-5255926

1 *Jaarboek 1945-1946.* (publ. Dec. 1946) [out of print] .... *f* 4,–
2 G.J.H. AMSHOFF, *Enumeration of the Herbarium Specimens of a Suriname Wood Collection.* (May 1948). Suppl. (Jan. 1950) .... *f* 7,50
3 A.M.W. MENNEGA, *Suriname Timbers.* (May 1948) .... *f* 15,–
4 *Jaarboek 1946-1948.* (June 1948) .... *f* 15,–
5 *Studies on the Fauna of Curaçao, Aruba, Bonaire and the Venezuelan Islands*, Vol. III, edited by P. WAGENAAR HUMMELINCK. (Nov. 1948) .... *f* 20,–
6 *Jaarboek 1948-1951.* (Nov. 1951) .... *f* 15,–
7 J.H. WESTERMANN, *Conservation in the Caribbean.* (Aug. 1952) .... *f* 15,–
8 *Studies on the Fauna of Curaçao and other Caribbean Islands*, Vol. IV. (June 1953) *f* 25,–
9 J.H. WESTERMANN, *Nature Preservation in the Caribbean.* (Dec. 1953) ...... *f* 15,–
10 *Studies on the Fauna of Curaçao* ..., Vol. V. (Oct. 1954) .... *f* 25,–
11 J.S. VEENENBOS, *A Soil and Land Capability Survey of St. Maarten, St. Eustatius, and Saba.* (May 1955) .... *f* 20,–
12 *Studies on the Fauna of Curaçao* ..., Vol. VI. (March 1955) .... *f* 25,–
13 F. HAVERSCHMIDT, *List of the Birds of Surinam.* (Nov. 1955) .... *f* 15,–
14 *Studies on the Fauna of Curaçao* ..., Vol. VII. (Jan. 1957) .... *f* 36,–
15 *Studies on the Flora of Curaçao* ..., Vol. I. (Sept. 1956). A.L. STOFFERS, *The Vegetation of the Netherlands Antilles* .... *f* 30,–
16 *Studies on the Fauna of Suriname and other Guyanas*, Vol. I. Edited by D.C. GEIJSKES & P. WAGENAAR HUMMELINCK. (March 1957) .... *f* 15,–
17 *Studies on the Fauna of Curaçao* ..., Vol. VIII. (July 1958) .... *f* 30,–
18 *Studies on the Fauna of Suriname* ..., Vol. II. (Feb. 1959) .... *f* 15,–
19 *Studies on the Fauna of Curaçao* ..., Vol. IX. (April 1959) .... *f* 15,–
20 *Studies on the Fauna of Suriname* ..., Vol. III. (Dec. 1959) .... *f* 25,–
21 *Studies on the Fauna of Curaçao* ..., Vol. X. (Feb. 1960) .... *f* 25,–
22 *Studies on the Fauna of Curaçao* ..., Vol. XI. (Dec. 1960) .... *f* 15,–
23 *Studies on the Fauna of Curaçao* ..., Vol. XII. (Dec. 1961) [Jan. 1962] ...... *f* 85,–
24 J.H. WESTERMANN & H. KIEL, *The Geology of Saba and St. Eustatius.* (Dec. 1961) [March 1962] .... *f* 35,–
25 *Flora of the Netherlands Antilles*, edited by A.L. STOFFERS, Vol. I (1), Pteridophyta. (Sept. 1962) .... *f* 12,–
26 *Studies on the Fauna of Curaçao* ..., Vol. XIII. (April 1962) .... *f* 20,–
27 *Studies on the Fauna of Suriname* ..., Vol. IV. (March 1963) .... *f* 20,–
28 *Studies on the Fauna of Suriname* ..., Vol. V. (Dec. 1962) [Jan. 1963] .... *f* 20,–
29 *Studies on the Fauna of Curaçao* ..., Vol. XIV. (April 1963) .... *f* 20,–
30 *Studies on the Fauna of Curaçao* ..., Vol. XV. (May 1963) .... *f* 20,–
31 *Studies on the Fauna of Curaçao* ..., Vol. XVI. (Sept. 1963) .... *f* 12,–

32 *Studies on the Fauna of Suriname* ..., Vol. VI. (Sept. 1963) . . . . *f* 10, –
33 *Studies on the Fauna of Curaçao* ..., Vol. XVII. (Oct. 1963) . . . . *f* 15, –
34 *Studies on the Fauna of Curaçao* ..., Vol. XVIII. (Dec. 1963) . . . . *f* 15, –
35 *Studies on the Fauna of Curaçao* ..., Vol. XIX. (Dec. 1963) . . . . *f* 15, –
36 *Flora of the Netherlands Antilles*, Vol. I (2) Monocotyledoneae. (Oct. 1963) . *f* 15, –
37 *Studies on the Fauna of Suriname* ..., Vol. VII. (April 1964) . . . . *f* 15, –
38 *Studies on the Fauna of Curaçao* ..., Vol. XX. (Sept. 1964) . . . . *f* 15, –
39 *Studies on the Fauna of Curaçao* ..., Vol. XXI. (Dec. 1964) . . . . *f* 20, –
40 *Studies on the Fauna of Curaçao* ..., Vol. XXII. (June 1965) . . . . *f* 20, –
41 JAN ROOTH, *The Flamingos on Bonaire.* (Dec. 1965) . . . . *f* 25, –
42 *Flora of the Netherlands Antilles*, Vol. II (1), Dicotyledoneae. (March 1966) . *f* 15, –
43 *Studies on the Fauna of Suriname* ..., Vol. VIII. (April 1966) . . . . *f* 25, –
44 *Studies on the Fauna of Suriname* ..., Vol. IX. (August 1966) . . . . *f* 20, –
45 *Studies on the Fauna of Curaçao* ..., Vol. XXIII. (Nov. 1966) . . . . *f* 30, –
46 *Studies on the Fauna of Curaçao* ..., Vol. XXIV. (March 1967) . . . . *f* 25, –
47 *Twintig Jaren Studiekring.* (Jan. 1968) . . . . *f* 15, –
48 *Studies on the Fauna of Suriname* ..., Vol. X. (April 1968) . . . . *f* 20, –
49 *Studies on the Fauna of Curaçao* ..., Vol. XXV. (June 1968) . . . . *f* 22, –
50 *Studies on the Fauna of Curaçao* ..., Vol. XXVI (April 1968) . . . . *f* 18, –
51 *Studies on the Fauna of Curaçao* ..., Vol. XXVII. (Sept. 1968) . . . . *f* 18, –
52 *Studies on the Flora of Curaçao* ..., Vol. II. (Nov. 1968), M. VROMAN, *The Marine Algal Vegetation of St. Martin, St. Eustatius and Saba* . . . . *f* 20, –
53 *Studies on the Fauna of Curaçao* ..., Vol. XXVIII. (Sept. 1969) . . . . *f* 20, –
54 *Studies on the Fauna of Curaçao* ..., Vol. XXIX. (June 1969) . . . . *f* 20, –
55 *Studies on the Fauna of Suriname* ..., Vol. XI. (May 1970) . . . . *f* 22, –
56 *Studies on the Fauna of Curaçao* ..., Vol. XXX. (Nov. 1969) . . . . *f* 18, –
57 *Studies on the Fauna of Curaçao* ..., Vol. XXXI. (Dec. 1969) . . . . *f* 22, –
58 *Studies on the Fauna of Curaçao* ..., Vol. XXXII. (March 1970) . . . . *f* 22, –
59 *Studies on the Fauna of Curaçao* ..., Vol. XXXIII. (Aug. 1970) . . . . *f* 20, –
60 *Studies on the Fauna of Suriname* ..., Vol. XII. (Nov. 1970) . . . . *f* 24, –
61 *Studies on the Fauna of Curaçao* ..., Vol. XXXIV. (Oct. 1970) . . . . *f* 24, –
62 *Studies on the Fauna of Curaçao* ..., Vol. XXXV. (March. 1971) . . . . *f* 22, –
63 *Studies on the Fauna of Curaçao* ..., Vol. XXXVI. (Sept. 1971) . . . . *f* 24, –
64 *Studies on the Fauna of Curaçao* ..., Vol. XXXVII. (Oct. 1971) . . . . *f* 25, –
65 *Studies on the Fauna of Curaçao* ..., Vol. XXXVIII. (Dec. 1971) . . . . *f* 25, –
66 *Studies on the Fauna of Curaçao* ..., Vol. I (Photomechanical reprint July 1971) *f* 20, –
67 *Studies on the Fauna of Curaçao* ..., Vol. II. (Photomechanical reprint July 1971) *f* 20, –
68 *Studies on the Fauna of Curaçao* ..., Vol. XXXIX. (March 1972) . . . . *f* 25, –
69 *Studies on the Fauna of Curaçao* ..., Vol. XL. (May 1972) . . . . *f* 28, –
70 D.J. BEETS, *Lithology and Stratigraphy of the Cretaceous and Danian Succession of Curaçao.* (May 1972) . . . . *f* 36, –
71 *Studies on the Fauna of Curaçao* ..., Vol. XLI. (July 1972) . . . . *f* 36, –
72 *Studies on the Fauna of Suriname* ..., Vol. XIII. (Nov. 1972) . . . . *f* 22, –
73 *Studies on the Fauna of Curaçao* ..., Vol. XLII. (Nov. 1972) . . . . *f* 48, –
74 *Studies on the Fauna of Curaçao* ..., Vol. XLIII. (May 1973) . . . . *f* 38, –
75 *Flora of the Netherlands Antilles*, Vol. III (1), Dicotyledoneae. (June 1973) . . *f* 15, –
76 *Studies on the Fauna of Curaçao* ..., Vol. XLIV. (Oct. 1973) . . . . *f* 40, –
77 *Studies on the Fauna of Curaçao* ..., Vol. XLV. (Feb. 1974) . . . . *f* 48, –

78 P.H. De Buisonje, *Neogene and Quaternary Geology of Aruba, Curaçao and Bonaire.* (March 1974) . . . . *f* 62, –
79 *Studies on the Fauna of Suriname* ..., Vol. XIV. (Aug. 1974) . . . . *f* 42, –
80 *Studies on the Fauna of Suriname* ..., Vol. XV. (March 1975) . . . . *f* 48, –
81 *Studies on the Fauna of Suriname* ..., Vol. XVI. (June 1975) . . . . *f* 88, –
82 *Studies on the Fauna of Curaçao* ..., Vol. XLVI. (Aug. 1975) . . . . *f* 48, –
83 *Studies on the Fauna of Curaçao* ..., Vol. XLVII. (Oct. 1975) . . . . *f* 48, –
84 *Studies on the Fauna of Curaçao* ..., Vol. XLVIII. (Nov. 1975) . . . . *f* 48, –
85 *Studies on the Fauna of Curaçao* ..., Vol. IL. (Aug. 1976) . . . . *f* 64, –
86 *Studies on the Fauna of Curaçao* ..., Vol. L. (Nov. 1976) . . . . *f* 48, –
87 *Studies on the Fauna of Curaçao* ..., Vol. LI. (Feb. 1977) . . . . *f* 50, –
88 *Studies on the Fauna of Curaçao* ..., Vol. LII. (June 1977) . . . . *f* 48, –
89 *Studies on the Fauna of Curaçao* ..., Vol. LIII. (July 1977) . . . . *f* 44, –
90 *Brokopondo Research Report, Suriname.* Part II: *Hydrobiology* (Dec. 1976) . . . . . *f* 48, –
91 *Studies on the Fauna of Curaçao* ..., Vol. LIV. (Oct. 1977) . . . . *f* 55, –
92 *Studies on the Fauna of Curaçao* ..., Vol. LV. (Nov. 1977) . . . . *f* 48, –
93 K.U. Kramer, *The Pteridophytes of Suriname.* (Aug. 1978) . . . . *f* 48, –
94 *Studies on the Fauna of Curaçao* ..., Vol. LVI. (Aug. 1978) . . . . *f* 48, –
95 P.G.E.F. Augustinus, *The changing shoreline of Surinam.* (Sept. 1978) . . . . . . *f* 65, –
96 *Studies on the Fauna of Curaçao* ..., Vol. LVII. (April 1979) . . . . *f* 90, –
97 P. Wagenaar Hummelinck, *Caves of the Netherlands Antilles.* (May 1979) . . *f* 27,50
98 *Studies on the Fauna of Curaçao* ..., Vol. LVIII. (Sep. 1979) . . . . *f* 30, –
99 *Studies on the Fauna of Curaçao* ..., Vol. LIX. (Oct. 1979) . . . . *f* 25, –
100 *Studies on the Fauna of Curaçao* ..., Vol. LX. (Oct. 1979) . . . . *f* 35, –
101 *Flora of the Netherlands Antilles*, Vol. III (2), Dicotyl. (Aug. 1979) . . . . *f* 30, –
102 *Flora of the Netherlands Antilles*, Vol. II (2), Dicotyl. (June 1980) . . . . *f* 30, –
103 *Studies on the Fauna of Curaçao* ..., Vol. LXI. (Oct. 1980) . . . . *f* 65, –
104 *Studies on the Fauna of Curaçao* ..., Vol. LXII. (Nov. 1980) . . . . *f* 70, –
105 J. Tacoma, *Studies on the physical anthropology of the N.A.: III-IV.* (Dec. 1980) *f* 12, –
106 *Flora of the Netherlands Antilles*, Vol. II (3), Dicotyl. (Feb. 1982) . . . . *f* 35, –
107 W.P. Nagelkerken, *Distribution and ecology of the Groupers and Snappers of the N.A.* (Oct. 1981) . . . . *f* 18, –
108 *Studies on the Fauna of Curaçao* ..., Vol. LXIII. (Dec. 1981) . . . . *f* 70, –
109 *De Surinaamsche Artz 1786-1788.* (Dec. 1981) . . . . *f* 15, –
110 *Studies on the Fauna of Curaçao* ..., Vol. LXIV. (April 1983) . . . . *f* 65, –
111 *Studies on the Fauna of Curaçao* ..., Vol. LXV. (Nov. 1983) . . . . *f* 60, –
112 *Studies on the Fauna of Curaçao* ..., Vol. LXVI. (June 1984) . . . . *f* 70, –
113 *Flora of the Netherlands Antilles*, Vol. III (3), Dicotyl. (Aug. 1984) . . . . *f* 65, –
114 *Studies on the Fauna of Curaçao* ..., Vol. LXVII (Dec. 1984) . . . . *f* 45, –
115 *Biotic and abiotic components of the landscapes of Saba.* (March 1985) . . . . *f* 15, –
116 *Biotic and abiotic components of the landscapes of St. Eustatius.* (March 1985) . . . *f* 15, –
117 Fleur C. van Duyl, *Atlas of the living reefs of Curaçao and Bonaire* (Nov. 1985) . . . . *f* 85, –
118 *Studies on the Fauna of Curaçao* ..., Vol. LXVIII. (Aug. 1986) . . . . *f* 65, –
119 G.Th. Klaver, *The Curaçao Lava Formation.* (Sept. 1987) . . . . *f* 30, –
120 J.B. Haviser, *Amerindian Cultural Geography on Curaçao.* (Nov. 1987) . . . . . . *f* 45, –
121 *Studies on the Fauna of Curaçao* ..., Vol. LXIX. (June 1988) . . . . *f* 85, –
122 *Studies on the Fauna of Curaçao* ..., Vol. LXX (Sept. 1988) . . . . *f* 75, –

123 *Studies in honour of Dr. Pieter Wagenaar Hummelinck.* (Dec. 1988) . . . . *f*100, –
124 *Brokopondo Research Report, Suriname* Part III: *The Vegetation.* (July 1989) . . . *f* 25, –
125 TH. E. WONG: *Revision of the stratigraphy of the coastal plain of Suriname.* (Aug. 1989) . . . . *f* 35, –
126 A. H. VERSTEEG, J. TACOMA & P. VAN DE VELDE, *Archaeological Investigations on Aruba; The Malmok Cemetry.* (Aug. 1990) . . . . *f* 15, –
127 J. TACOMA, D.C. GEIJSKES, G. J. R. MAAT & G. N. VAN VARK, *On 'Amazonidi', Precolumbian Skeletal Remains and Associated Archaeology from Suriname* (Sept. 1991) . . . . *f* 45, –
128 G. J. VAN KLINKEN, *Dating and dietary reconstruction by isotopic analysis of amino acids in fossil bone collagen – with special reference to the Caribbean.* (March 1991). . . . *f* 45, –
129 C. N. DUBELAAR, *Bibliography of South American and Antillean Petroglyphs.* (July 1991) . . . . *f* 35, –
130 P. STIENSTRA, *Sedimentary Petrology, Origin and Mining History of the Phosphate Rocks of Klein Curaçao, Curaçao and Aruba, Netherlands West Indies.* (Dec. 1991) *f* 60, –
131 AAD H. VERSTEEG & KEES SCHINKEL, *The Archaeology of St. Eustatius: The Golden Rock Site.* (Dec. 1992) . . . . *f* 35, –
132 *Studies on the Natural History of the Caribbean Region*, Vol. LXXI. (Dec. 1992) *f* 45, –
133 BRUCE W. FOUKE,*Depositions, Diagenesis and Dolomitization of Neogene Seroe Domi Formation Coral Reef Limestones on Curaçao, Netherlands Antilles* (March 1994) . . . . *f* 55, –